T0261433

MOLECULAR MACHINES

Molecular Machines

A Materials Science Approach

Giovanni Zocchi

PRINCETON UNIVERSITY PRESS

PRINCETON AND OXFORD

Copyright © 2018 by Princeton University Press

Published by Princeton University Press,
41 William Street, Princeton, New Jersey 08540

In the United Kingdom: Princeton University Press,
6 Oxford Street, Woodstock, Oxfordshire, OX20 1TR

All Rights Reserved

LCCN 2017958475
ISBN 978-0-691-17386-3

British Library Cataloging-in-Publication Data is available

This book has been composed in Adobe Text Pro and Gotham

Printed on acid-free paper. ∞

press.princeton.edu

Typeset by Nova Techset Pvt Ltd, Bangalore, India

Printed in the United States of America

10   9   8   7   6   5   4   3   2   1

CONTENTS

Preface   vii

1   Brownian Motion   1

1.1   *Random Walk*   1

1.2   *Polymer as a Simple Random Walk*   4

1.3   *Direct Calculation of $p(\vec{R})$*   5

1.4   *The Langevin Approach*   8

1.5   *Correlation Functions*   12

1.6   *Barrier Crossing*   16

1.7   *What is Equilibrium?*   22

2   Statics of DNA Deformations   25

2.1   *Introduction*   25

2.2   *DNA Melting*   29

2.3   *Zipper Model*   34

2.4   *Experimental Melting Curves*   36

2.5   *Base Pairing and Base Stacking as Separate*
       Degrees of Freedom   42

2.6   *Hamiltonian Formulation of the Zipper Model*   46

2.7   *2×2 Model: Cooperativity from*
       Local Rules   47

2.8   *Nearest Neighbor Model*   52

2.9   *Connection to Nonlinear Dynamics*   55

2.10   *Linear and Nonlinear Elasticity of DNA*   56

2.11   *Bending Modulus and Persistence Length*   56

2.12   *Measurements of DNA Elasticity: Long Molecules*   60

2.13   *Measurements of DNA Elasticity: Short Molecules*   65

2.14   *The Euler Instability*   68

2.15   *The DNA Yield Transition*   72

3 Kinematics of Enzyme Action   81

    3.1   *Introduction*   81

    3.2   *Michaelis–Menten Kinetics*   82

    3.3   *The Method of the DNA Springs*   89

    3.4   *Force and Elastic Energy in the Enzyme—DNA Chimeras*   97

    3.5   *Injection of Elastic Energy vs. Activity Modulation*   106

    3.6   *Connection to Nonlinear Dynamics: Two Coupled Nonlinear Springs*   116

4 Dynamics of Enzyme Action   122

    4.1   *Introduction*   122

    4.2   *Enzymes are Viscoelastic*   125

    4.3   *Nonlinearity of the Enzyme's Mechanics*   125

    4.4   *Timescales*   128

    4.5   *Enzymatic Cycle and Viscoelasticity: Motors*   129

    4.6   *Internal Dissipation*   137

    4.7   *Origin of the Restoring Force g*   138

    4.8   *Models Based on Chemical Kinetics (Fisher and Kolomeisky, 1999)*   139

    4.9   *Different Levels of Microscopic Description*   143

    4.10   *Connection to Information Flow*   147

    4.11   *Normal Mode Analysis*   149

    4.12   *Many States of the Folded Protein: Spectroscopy*   153

    4.13   *Interesting Topics in Nonequilibrium Thermodynamics Relating to Enzyme Dynamics*   158

Bibliography   165

    Chapter 1: Brownian Motion   165

    Chapter 2: Statics of DNA Deformations   165

    Chapter 3: Kinematics of Enzyme Action   167

    Chapter 4: Dynamics of Enzyme Action   169

Index   173

The book presents a traditional subject in biophysics—conformational dynamics of biological macromolecules—from a somewhat unorthodox perspective. Biology-oriented research in this interdisciplinary area is dominated by structural analysis and thinking (the "structure–function paradigm"). Physics-oriented research has focused heavily on the development of techniques ("single molecule biophysics") and less on concepts. An exception is the early single molecule work on molecular motors (and supporting theoretical developments), which started to address squarely the fundamentals of dynamics.

Our own angle is a generalization of that approach, where the important quantities are dynamic: forces and deformations, stress and strain, to a larger class of enzymes, and indeed other macromolecules, including DNA. One can pose the question of "force–velocity curves" even for enzymes that are not motors. Similarly, one can pose the question of the deformability of enzymes beyond the crystallographically documented conformational transitions.

Starting from certain recent experimental developments, we build a way of thinking about conformational motion (the molecular machine aspect of living systems) that is materials science oriented (as opposed to structural-biology oriented). We find this approach conducive to designing innovative experiments in the field. The book is, therefore, mainly directed to the experimentalist who wishes to push the boundaries of materials science. However, it is a book about concepts, not techniques. Always the concept is illustrated through a relatively simple model, and often the purpose of the model is not so much to fit a particular set of data, as it is a vehicle for thinking about the system and asking questions about the system.

The focus on dynamical quantities—forces, deformations, elastic energies—allows some fundamental unity of the subject matter to emerge that is otherwise not apparent. One such "universal" process is softening transitions. Large conformational motion accesses mechanical regimes beyond linear elasticity that seem invariably characterized by a local softening of the folded structure. In the book, such reversible yield transitions, in their static and dynamic forms, are the guiding thread for chapter 2 on DNA mechanics and chapter 4 on enzyme mechanics (and, to a lesser extent, chapter 3).

Some topics in the book (still pertaining to conformational changes) are treated in a more traditional manner, for instance DNA melting. This is a subject that has received a lot of attention (perhaps too much attention) from physicists (alas, including the author) in the context of phase transitions and force-induced unzipping. However, here too our focus is slightly unorthodox in that it is *not* on the thermodynamic limit of long molecules, but rather on short molecules as a clean experimental system and nanoscience tool.

In summary, through certain recent experimental developments, the subject of conformational transitions (central to the molecular machines aspect of life) is acquiring a materials science basis that is not yet reflected in the secondary literature. The physics-oriented presentations of the subject are mostly informed by concepts of the energy landscape which were really developed to describe the folding–unfolding transition (another favorite among physicists) but do not provide an incisive description of the dynamics of conformational transitions of the folded state, which is quite a different, and maybe simpler, problem. This book attempts to formulate the subject in terms more resonant with materials scientists, with the intent of opening this area of research to a wider nanoscience community, especially experimentalists.

A word about the more practical aspects of the format. I thought of addressing the book to the graduate students first coming to work in my group. One question is always, "What should I read to get started?" With this work, I mean to provide part of the answer. I assume a good grounding in equilibrium statistical mechanics. Building on that understanding, the book is also to some extent a collection of exercises in statistical physics. One can practice the transfer matrix method on the example of the melting of the DNA double helix, instead of the more usual ferromagnetic transition. I think of the subject matter of the book as part of condensed matter physics, so I sought to make the material deliverable in lecture form to a class of physics students. I like my lectures to take place at the blackboard, and I wrote the book with that in mind; calculations are presented step by step. I opted not to encumber the text with a reference for each statement; this is normal practice for a textbook, while the opposite is true for a monograph. My book lies somewhat in between. There are still many references for each chapter, but most are not referred to in the text. Exceptions are when figures or whole calculations are directly taken from one reference, which is then acknowledged explicitly. Consistent with this approach, scant attention is paid to the history of the subject and questions of priority. This book is about the science, not the scientists.

I did not include a chapter on biomolecular structures. To those unfamiliar with, say, the structure of DNA, I say first that they should be familiar with it, and second that they are only one short typeset away from an excellent

Wikipedia article on the topic. The same goes for any biological term in the book that may be unfamiliar to the physicist. On the other hand, I did include a chapter (chapter 1) on Brownian motion. There are two reasons: One is that while it is safe to assume knowledge of equilibrium statistical mechanics in first- or second-year physics graduate students, the opposite is true when it comes to Brownian motion. This topic partakes of both equilibrium and nonequilibrium statistical mechanics; unless it is appreciated, the whole subject matter of the book has little conceptual underpinning.

Finally, a note on notation. In all the formulas, we use reduced units where the Boltzmann constant $= 1$. Therefore the temperature T has dimensions of energy and the entropy S is dimensionless. On the other hand, we use the notation $k_B T$ as an energy unit corresponding to the thermal energy at room temperature: $1\,k_B T = 4.2\,\mathrm{pN\,nm} = 25\,\mathrm{meV}$, approximately. While this usage is nonstandard in a book, it has become the common "spoken" usage in the field.

Acknowledgments

This book owes its existence to the author's students. My understanding of the subject matter, and the approach adopted in the book, were developed with my students and through their work. In addition, the work in question would not have happened without the financial support of the National Science Foundation, DMR.

At Princeton University Press, I was fortunate to work with Eric Henney, who believed in the project and oversaw it from the start, as well as the rest of the staff who carried it through with a high degree of professionalism.

Last but not least I wish to thank my family, which includes individuals of the human, canine, and feline persuasion.

I dedicate this book to the memory of my father, Sergio Zocchi.

MOLECULAR MACHINES

1

Brownian Motion

The phenomenon of Brownian motion connects equilibrium and non-equilibrium statistical mechanics. It connects diffusion—a nonequilibrium phenomenon—with thermal fluctuations—an equilibrium concept. More precisely, diffusion with a net flow of particles, driven by a concentration gradient, pertains to a nonequilibrium system, since there is a net current. Without a concentration gradient, the system is macroscopically in equilibrium, but each individual particle undergoes self-diffusion just the same. In this sense, Brownian motion is at the border of equilibrium and nonequilibrium statistical mechanics. Understanding Brownian motion led Einstein, in one of his famous 1905 papers, to a form of the fluctuation–dissipation theorem. Here we give an introduction to the main ideas.

1.1 Random Walk

The simplest model of Brownian motion is a random walk on a lattice. It is the following process: a particle starts at a lattice site, and makes random steps to the neighboring sites. For example, we may visualize a square lattice in 2-D (figure 1.1). With equal probability, the particle steps up, down, right, or left. After a time interval τ_0 the process is repeated. Thus time is connected to the number of steps N by

$$t = N\tau_0. \tag{1.1}$$

The ith step is specified by a vector $\vec{r}_i$ (there are only 4 such vectors for a square lattice in 2-D); a particular realization of the random walk of N steps is the set of vectors

$$\text{random walk:} \quad \{\vec{r}_i : i = 1, 2, \ldots, N\}, \quad |\vec{r}_i| = \ell \quad \forall i, \tag{1.2}$$

FIGURE 1.1. Square lattice and the vector $\vec{r}_i$ representing the ith step of a random walk.

where ℓ is the step size (the lattice spacing). The end-to-end distance of the walk (the displacement of the particle after N steps) is

$$\vec{R} = \sum_{i=1}^{N} \vec{r}_i. \tag{1.3}$$

The $\vec{r}_i$'s are random variables, and so is $\vec{R}$. Now we ask what the *typical* displacement of the particle is, that is, we want ensemble averages. Because of the symmetry, $\langle \vec{r}_i \rangle = \vec{0}$ and so

$$\langle \vec{R} \rangle = \left\langle \sum_i \vec{r}_i \right\rangle = \sum_i \langle \vec{r}_i \rangle = \vec{0}; \tag{1.4}$$

however, the second moment is not zero:

$$\langle R^2 \rangle = \left\langle \left| \sum_{i=1}^{N} \vec{r}_i \right|^2 \right\rangle = \sum_{i,j} \langle \vec{r}_i \cdot \vec{r}_j \rangle, \tag{1.5}$$

where we use the notation $R^2 \equiv |\vec{R}|^2$. Since different steps are uncorrelated (independent), we have

$$\langle \vec{r}_i \cdot \vec{r}_j \rangle = \ell^2 \delta_{ij} \tag{1.6}$$

and therefore

$$\langle R^2 \rangle = N\ell^2. \tag{1.7}$$

The mean square displacement is $\propto N$, or the rms displacement is $\propto \sqrt{N}$. For the Brownian particle, since $N = t/\tau_0$, we have

$$\langle R^2 \rangle = \frac{\ell^2}{\tau_0} t. \tag{1.8}$$

The parameter ℓ^2/τ_0 has dimensions of a diffusion constant (length^2/time). Indeed, we recall that for the process of diffusion, described by the diffusion equation

$$\frac{\partial c(\vec{x}, t)}{\partial t} - D\nabla^2 c = 0, \tag{1.9}$$

where $c(\vec{x}, t)$ is the concentration of particles (number of particles per unit volume) and D is the diffusion constant, starting with a δ-function concentration at time 0, the width of the distribution c spreads in time according to

$$\langle R^2 \rangle^{1/2} \sim \sqrt{Dt}, \tag{1.10}$$

so we identify

$$D \sim \frac{\ell^2}{\tau_0}. \tag{1.11}$$

The process of the random walk is the process of diffusion seen from the viewpoint of the individual particle.

Besides calculating individual moments, for the simple random walk problem it is not difficult to calculate the whole probability distribution of the end-to-end distance. Even off-lattice, we can write down off-hand,

$$p(\vec{R}) \propto \exp\left(-\frac{3R^2}{2\langle R^2 \rangle}\right) \tag{1.12}$$

by applying the central limit theorem. It states that the sum of (a large number of) independent random variables is distributed as a Gaussian. Then (1.12) follows from (1.3). Obviously, a precise mathematical formulation needs a few more specifications, but in our case the random variables in question (the $\vec{r}_i$'s) are not only independent but also identically distributed; the theorem applies in a strong form, and we do not worry about details. Instead, we give below a "proof" of this result in a particular case.

The factor of 3 in the exponent in (1.12) comes from the dimensionality (3-D): in 1 dimension the probability distribution is

$$p(x) \propto \exp\left(-\frac{x^2}{2\langle x^2 \rangle}\right), \tag{1.13}$$

and because the 3 directions are independent,

$$p(x, y, z) \propto \exp\left(-\frac{x^2}{2\langle x^2 \rangle}\right) \exp\left(-\frac{y^2}{2\langle y^2 \rangle}\right) \exp\left(-\frac{z^2}{2\langle z^2 \rangle}\right). \tag{1.14}$$

Furthermore, $\langle x^2 \rangle = \langle y^2 \rangle = \langle z^2 \rangle = \langle R^2 \rangle / 3$ (since $R^2 = x^2 + y^2 + z^2$); therefore from (1.14) we obtain (1.12). Using the result (1.7), we can write for the probability distribution of the end-to-end distance,

$$p(\vec{R}) \propto \exp\left(-\frac{3R^2}{2N\ell^2}\right), \tag{1.15}$$

which refers to a random walk of N steps in 3-D, with $N \gg 1$. The normalization constant is given by the requirement

$$\int_0^\infty dR \, 4\pi R^2 p(\vec{R}) = 1, \tag{1.16}$$

leading to

$$p(\vec{R}) = \left(\frac{3}{2\pi N \ell^2}\right)^{3/2} \exp\left(-\frac{3R^2}{2N\ell^2}\right). \tag{1.17}$$

Note that this is the probability distribution for the *vector* $\vec{R}$; it is a Gaussian centered at $\vec{R} = \vec{0}$. This result also says that the most probable endpoint of the walk is back where it started.

On the other hand, if we ask for the probability that the walk ends up at some *distance* $R \equiv |\vec{R}|$ from where it started, we have

$$p(R) = 4\pi R^2 p(\vec{R}) \propto R^2 \exp\left(-\frac{3R^2}{2N\ell^2}\right), \tag{1.18}$$

which is zero for $R = 0$ and maximum for $R = \sqrt{2N/3}\,\ell$.

In conclusion, the simple random walk can be solved exactly, and the statistics is Gaussian for large N. The adjective "simple" indicates that there are more complicated problems, derived from this simple one, which are also of interest. For instance, we could add boundary conditions to the process, say a confining wall. We could add a field that biases the probabilities of the different steps. Most important, we could require the walk to be self-avoiding, that is, a trajectory is not allowed to cross itself. All these additions correspond to important physical situations.

1.2 Polymer as a Simple Random Walk

The simplest model for a long, flexible polymer in a good solvent (so that the polymer is not collapsed) is to visualize the different conformations of the chain as realizations of a random walk of N steps of size ℓ_K. The parameter ℓ_K represents the length scale over which the polymer chain is substantially flexible (i.e., at shorter lengths the polymer is rigid). Therefore, due to thermal fluctuations, the directions of the chain at two points separated by ℓ_K or more are essentially uncorrelated, like successive steps in the random walk. In polymer physics, ℓ_K is called the Kuhn length ($\ell_K = 2\ell_p$; ℓ_p is called the persistence length), and it is a parameter characteristic of the specific polymer. Here, N is the number of Kuhn lengths in the polymer chain, that is, $N\ell_K$ is the contour length of the chain. We consider the number of configurations of the chain, $\Gamma(\vec{R})$, which correspond to an end-to-end distance $\vec{R}$ (figure 1.2); we further assume that all configurations of the chain have the same energy (which amounts to considering only entropic effects).

Then

$$\Gamma(\vec{R}) \propto p(\vec{R}) \propto \exp\left(-\frac{3R^2}{2N\ell_K^2}\right), \quad (1.19)$$

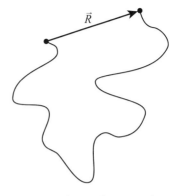

using (1.15). Since all states are equiprobable in this simplest model, the entropy of the chain as a function of the end-to-end distance $\vec{R}$ is

$$S(\vec{R}) = \ln \Gamma(\vec{R}) = \ln p(\vec{R}) + \text{const.}, \quad (1.20)$$

and from (1.19),

$$S(\vec{R}) = -\frac{3R^2}{2N\ell_K^2} + \text{const.} \quad (1.21)$$

FIGURE 1.2. One configuration of a flexible polymer chain with end-to-end distance $\vec{R}$.

The free energy as a function of $\vec{R}$ is therefore, apart from a constant (R-independent) term,

$$F(\vec{R}) = E - TS = T\frac{3R^2}{2N\ell_K^2}, \quad (1.22)$$

because we suppose E independent of $\vec{R}$.

Since the free energy (1.22) increases with R (the distance between the endpoints), it requires a force to keep the endpoints of the chain separated by some distance. That force, which is purely entropic (since there is no energy change with R) is

$$\text{force} = -\frac{\partial F}{\partial R} = -\frac{3T}{N\ell_K^2}R, \quad (1.23)$$

proportional to the distance between the endpoints. The polymer chain acts as an entropic spring! The spring constant ($3T/N\ell_K^2$) *increases* with temperature, signalling that this form of elasticity is physically quite different from the usual elasticity, which originates in energy differences, not entropy differences. The reason why the entropy of the chain decreases with increasing end-to-end distance (eq. (1.21)) is depicted in figure 1.3. Formulas (1.22) and (1.23) are valid for small elongations $R \ll N\ell_K$; otherwise, the statistics of the chain is not Gaussian.

1.3 Direct Calculation of $p(\vec{R})$

We consider a random walk of N steps in 1-D, with steps of size 1. Let

$$\begin{cases} r = \text{number of steps to the right,} \\ \ell = \text{number of steps to the left;} \end{cases} \quad (1.24)$$

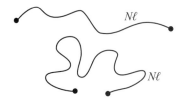

$N\ell$

$N\ell$

FIGURE 1.3. A large end-to-end distance is realized by a small number of conformations, while a small end-to-end distance can be realized by many more conformations of the chain. The value $N\ell$ is the contour length of the chain.

then

$$\begin{cases} r + \ell = N, \\ r - \ell = x \end{cases} \Rightarrow \begin{cases} r = \frac{N+x}{2}, \\ \ell = \frac{N-x}{2}, \end{cases} \tag{1.25}$$

where N is the number of steps, and x is the end-to-end distance. The number of walks of size N with r steps to the right is, exactly,

$$N(r) = \frac{N!}{r!\ell!} = \frac{N!}{r!(N-r)!} = \binom{N}{r}. \tag{1.26}$$

We can reason as follows. Start with one realization of a walk with r steps to the right. We can generate all other realizations by permutations of all the steps, and there are $N!$ such. However, we have overcounted by the following factors: Permutations among the subset of steps to the right produce the *same* walk. There are $r!$ such. Similarly for the steps to the left. Hence (1.26). Using (1.25) and (1.26), the number of walks with end-to-end distance x is

$$N(x) = \frac{N!}{\left(\frac{N+x}{2}\right)!\left(\frac{N-x}{2}\right)!}. \tag{1.27}$$

In this formula, x is an integer because we consider steps of size 1. By inspection we see that $N(x)$ is maximum for $x = 0$: the largest number of walks occurs for coming back to the same place. We transform (1.27) using the approximation

$$\ln n! \approx n\ln n - n \quad \Leftrightarrow \quad n! \approx \left(\frac{n}{e}\right)^n \quad (n \gg 1), \tag{1.28}$$

valid for large n. We obtain

$$N(x) \approx \frac{(N/e)^N}{\left(\frac{N+x}{2e}\right)^{(N+x)/2}\left(\frac{N-x}{2e}\right)^{(N-x)/2}}$$

$$= \frac{2^N}{\left(1 + \frac{x}{N}\right)^{(N+x)/2}\left(1 - \frac{x}{N}\right)^{(N-x)/2}}, \tag{1.29}$$

the last step after a little algebra. Note that to write down the approximate form (1.29) starting from the exact result (1.27), we require $x \ll N$ (and of course $N \gg 1$). Taking the log of (1.29),

$$\ln N(x) = N \ln 2 - \frac{N+x}{2} \ln\left(1 + \frac{x}{N}\right) - \frac{N-x}{2} \ln\left(1 - \frac{x}{N}\right). \tag{1.30}$$

Since x/N is supposed a small parameter, we use the expansion

$$\ln(1 + \varepsilon) \approx \varepsilon - \frac{\varepsilon^2}{2} \quad (\varepsilon \ll 1). \tag{1.31}$$

It is necessary to expand to second order in ε to keep orders consistently, as we see from the next equation, where we use (1.31) in (1.30):

$$\ln N(x) \approx N \ln 2 - \frac{N+x}{2}\left(\frac{x}{N} - \frac{x^2}{2N^2}\right) - \frac{N-x}{2}\left(-\frac{x}{N} - \frac{x^2}{2N^2}\right)$$

$$= N \ln 2 - \frac{x^2}{2N} = N\left[\ln 2 - \frac{x^2}{2N^2}\right], \tag{1.32}$$

correct to order $(x/N)^2$. Finally we obtain the Gaussian approximation for the number of walks with end-to-end distance x:

$$N(x) \approx 2^N \exp\left(-\frac{x^2}{2N}\right) \quad (x \ll N). \tag{1.33}$$

The prefactor in (1.33) is a casualty of approximation (1.28), since 2^N is the total number of walks. We find a better normalization by noting that

$$\int_{-\infty}^{+\infty} dx \, \exp\left(-\frac{x^2}{2N}\right) = \sqrt{2\pi N} \tag{1.34}$$

and write

$$N(x) = \frac{2^N}{(2\pi N)^{1/2}} \exp\left(-\frac{x^2}{2N}\right). \tag{1.35}$$

The probability of the end-to-end distance x is found by dividing the number of walks (1.35) by the total number of walks (since all walks are equiprobable):

$$p(x) = \frac{N(x)}{2^N} = \frac{1}{(2\pi N)^{1/2}} \exp\left(-\frac{x^2}{2N}\right) \quad \text{(1-D)}. \tag{1.36}$$

The factor 2^N in (1.35) comes from the coordination number of the lattice (the number of nearest neighbors), which is 2 in 1-D. However, the probability distribution (1.36) is independent of the lattice. In 3-D, with a coordination number C (i.e., the total number of walks is C^N), the formulas above become

$$N(\vec{x}) = \frac{C^N}{(2\pi N)^{3/2}} \exp\left(-\frac{3|\vec{x}|^2}{2N}\right), \tag{1.37}$$

while the probability is independent of the lattice (and therefore all the moments too):

$$p(x) = \frac{1}{(2\pi N)^{3/2}} \exp\left(-\frac{3|\vec{x}|^2}{2N}\right) \quad \text{(3-D)}. \tag{1.38}$$

We can put back the step size ℓ in the formulas above through the substitution $|\vec{x}| \to |\vec{x}|/\ell$. Thus we have seen directly that the statistics of the random walk is Gaussian.

1.4 The Langevin Approach

We consider again a Brownian particle, that is, a free particle of mass m in a fluid, with no average force acting. We write the following equation of motion for the particle:

$$m\frac{d\vec{v}}{dt} = -\gamma\vec{v} + \vec{\Gamma}(t), \tag{1.39}$$

where $\vec{v}(t)$ is the velocity of the particle, γ is the viscous damping coefficient, and $\vec{\Gamma}(t)$ is the random force on the particle originating from molecular collisions, with

$$\langle\vec{\Gamma}(t)\rangle = \vec{0}. \tag{1.40}$$

Equation (1.39) is a stochastic differential equation, because $\vec{\Gamma}(t)$ is a random variable. At one level, (1.39) is simply Newton's second law: the right-hand side is the force on the particle, consisting of a drag term $(-\gamma\vec{v})$ and an "external" force $(\vec{\Gamma})$ due to molecular collisions. On the other hand, it is not obvious that one can divide the force acting on the particle in this way. The drag *also* comes from molecular collisions. Equation (1.39) is a statement that mechanical equilibrium is established instantaneously. The two force terms on the right-hand side of (1.39) ($[\gamma v] = [\Gamma]$ = force) operate in general over very different timescales: $\vec{\Gamma}(t)$ varies over fast timescales of order τ_0 (the time between steps in the random walk problem), while $\gamma\vec{v}$ varies over slower characteristic timescales of order τ, defined below. Formulation (1.39) is generally useful when $\tau_0 \ll \tau$, which holds true, for example, for a micron-size Brownian particle in water.

The dissipation parameter γ is related to the mobility of the particle in the fluid: if we apply a constant external force $\vec{F}_{ext}$, the equation of motion (1.39) becomes

$$m\frac{d\vec{v}}{dt} = -\gamma\vec{v} + \vec{\Gamma}(t) + \vec{F}_{ext}, \tag{1.41}$$

and taking ensemble averages (using (1.40)),

$$m\frac{d}{dt}\langle\vec{v}\rangle = -\gamma\langle\vec{v}\rangle + \vec{F}_{ext}. \tag{1.42}$$

We have used $\langle d\vec{v}/dt \rangle = (d/dt)\langle \vec{v} \rangle$, which is formally correct if one interprets $\langle\ \rangle$ strictly as an ensemble average, but is not so obvious if one instead interprets $\langle\ \rangle$ as a running time average over the particle's trajectory.

A particular solution of (1.42) is the steady state solution $(d/dt)\langle \vec{v} \rangle = \vec{0}$:

$$\gamma\langle \vec{v} \rangle = \vec{F}_{ext} \quad \Rightarrow \quad \langle \vec{v} \rangle = \frac{1}{\gamma}\vec{F}_{ext} = \mu\vec{F}_{ext}, \tag{1.43}$$

where $\mu = 1/\gamma$ is the mobility of the particle. The solution of the homogeneous equation

$$\frac{d}{dt}\langle \vec{v} \rangle + \frac{\gamma}{m}\langle \vec{v} \rangle = \vec{0} \tag{1.44}$$

is

$$\langle \vec{v} \rangle = \langle \vec{v}(0) \rangle e^{-t/\tau}, \quad \tau = \frac{m}{\gamma}, \tag{1.45}$$

and introduces the relaxation time τ for the "smooth" part of the velocity of the particle. Thus the general solution of (1.42) is

$$\langle \vec{v}(t) \rangle = \langle \vec{v}(0) \rangle e^{-t/\tau} + \mu\vec{F}_{ext}, \tag{1.46}$$

where

$$\mu = \frac{1}{\gamma} \tag{1.47}$$

is the mobility. The solution (1.46) consists of a transient plus a steady drift. The particle "forgets" initial conditions after a time of order τ. Let us take the case of a $1\,\mu$m sphere in water: for γ we use the hydrodynamic result (Stokes formula) $\gamma = 6\pi R\eta$, where R is the radius of the sphere and η is the viscosity of water; in cgs units, $R = 0.5 \times 10^{-4}$, $\eta \approx 10^{-2}$. Assuming a density $\rho = 1$ for the particle, we find from (1.45),

$$\tau = \frac{m}{\gamma} = \frac{(4/3)\pi R^3\rho}{6\pi R\eta} = \frac{2}{9}\frac{R^2\rho}{\eta} \approx 0.2 \times \frac{(0.5 \times 10^{-4})^2}{10^{-2}}\,\text{s} \tag{1.48}$$

$$\approx 4 \times 10^{-8}\,\text{s} = 40\,\text{ns},$$

still large compared to the molecular collision time, which is $\tau_0 < 1$ ps. However, we see from (1.48) that $\tau \propto R^2$, so for a 1 nm sphere these two timescales would be of the same order.

Now we calculate the mean square displacement of the particle given by the Langevin equation (1.39), which we rewrite as

$$\frac{d\vec{v}}{dt} = -\frac{1}{\tau}\vec{v} + \frac{1}{m}\vec{\Gamma}(t), \tag{1.49}$$

with $\tau = m/\gamma$. Take the scalar product of (1.49) with $\vec{x}$, ensemble average $\langle\ \rangle$, and note that

$$\frac{d\vec{v}}{dt}\cdot\vec{x} = \frac{1}{2}\frac{d^2}{dt^2}x^2 - v^2, \qquad \vec{v}\cdot\vec{x} = \frac{1}{2}\frac{d}{dt}x^2, \tag{1.50}$$

and also

$$\langle\vec{\Gamma}(t)\cdot\vec{x}(t)\rangle = \langle\vec{\Gamma}(t)\rangle\cdot\langle\vec{x}(t)\rangle = 0 \tag{1.51}$$

because $\vec{\Gamma}(t)$ and $\vec{x}(t)$ are uncorrelated. The result is

$$\frac{d^2}{dt^2}\langle x^2\rangle + \frac{1}{\tau}\frac{d}{dt}\langle x^2\rangle = 2\langle v^2\rangle. \tag{1.52}$$

By equipartition, we have

$$\frac{1}{2}m\langle v^2\rangle = \frac{3}{2}T \tag{1.53}$$

at equilibrium, so we write (1.52) as

$$\frac{d^2}{dt^2}\langle x^2\rangle + \frac{1}{\tau}\frac{d}{dt}\langle x^2\rangle = \frac{6T}{m}. \tag{1.54}$$

The two solutions of the corresponding homogeneous equation are of the form

$$\langle x^2\rangle \sim e^{at}, \quad a = 0 \text{ or } a = -\frac{1}{\tau}. \tag{1.55}$$

A particular solution of (1.54) is

$$\frac{d}{dt}\langle x^2\rangle = \frac{6T}{m}\tau \quad \Rightarrow \quad \langle x^2\rangle = \frac{6T}{m}\tau t, \tag{1.56}$$

so the general solution of (1.54) is

$$\langle x^2\rangle = \frac{6T}{m}\tau t + Ae^{-t/\tau} + B, \tag{1.57}$$

where A, B are arbitrary constants. With the boundary condition $\langle x^2\rangle = 0$ at $t = 0$, we have $B = -A$ and therefore

$$\langle x^2(t)\rangle = \frac{6T}{m}\tau t + A(e^{-t/\tau} - 1). \tag{1.58}$$

For long times $t \gg \tau$, we obtain diffusive behavior:

$$\langle x^2(t)\rangle \approx \frac{6T}{m}\tau t \propto t. \tag{1.59}$$

For short times $t \ll \tau$, we should have ballistic behavior ($\langle x^2\rangle \propto t^2$), and we can use this fact to find the constant A. From (1.58), for short times,

$$\langle x^2\rangle \approx \frac{6T\tau}{m}t - A\frac{t}{\tau} + \frac{1}{2}A\left(\frac{t}{\tau}\right)^2, \tag{1.60}$$

and we must have

$$A = \frac{6T\tau^2}{m} \tag{1.61}$$

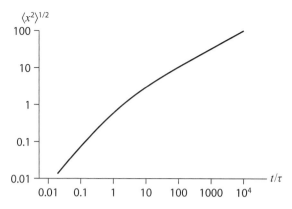

FIGURE 1.4. Log-log plot of $\langle x^2 \rangle^{1/2}$ (in units of $\sqrt{6T\tau^2/m}$) vs. t/τ according to eq. (1.63).

so that

$$\langle x^2 \rangle = \frac{3T}{m}t^2 = \langle v^2 \rangle t^2, \tag{1.62}$$

consistent with ballistic motion. Finally,

$$\langle x^2 \rangle = \frac{6T\tau^2}{m}\left[\frac{t}{\tau} + (e^{-t/\tau} - 1)\right]. \tag{1.63}$$

In figure 1.4 we plot $\langle x^2 \rangle^{1/2}$ vs. t according to (1.63): the graph starts linear, and goes over to a square root. We now examine the relation to the diffusion equation in more quantitative detail. The diffusion equation arises from the conservation law for the number of particles:

$$\frac{\partial n(\vec{x}, t)}{\partial t} + \vec{\nabla} \cdot \vec{j} = 0, \tag{1.64}$$

where n is the density of particles (number of particles per unit volume) and $\vec{j}$ is the particles' current. If the current is proportional to the concentration gradient,

$$\vec{j}(\vec{x}, t) = -D\vec{\nabla}n, \tag{1.65}$$

we have the diffusion equation

$$\frac{\partial n}{\partial t} - D\nabla^2 n(\vec{x}, t) = 0, \tag{1.66}$$

where D is the diffusion constant, with dimensions $[D] = \ell^2/t$, as we see from (1.66). With initial condition

$$n(\vec{x}, 0) = N\delta(\vec{x}) \tag{1.67}$$

and free boundaries (N is the total number of particles), the solution of (1.66) is

$$n(\vec{x}, t) = \frac{N}{(4\pi Dt)^{3/2}} \exp\left(-\frac{x^2}{4Dt}\right). \tag{1.68}$$

The probability of finding a particle at $(\vec{x}, t)$ is $p(\vec{x}, t) = n(\vec{x}, t)/N$, so now we can calculate all the moments. We find

$$\langle \vec{x}(t) \rangle = \frac{1}{N} \int d^3x \, n(\vec{x}, t)\vec{x} = \vec{0}, \tag{1.69}$$

whereas

$$\langle x^2(t) \rangle = \frac{1}{N} \int d^3x \, n(\vec{x}, t)x^2 = \frac{1}{(4\pi Dt)^{3/2}} \int_0^\infty dx \, 4\pi x^4 \exp\left(-\frac{x^2}{4Dt}\right), \tag{1.70}$$

where we have put the integral in spherical coordinates. With the substitution $x/\sqrt{4Dt} = u$, the right-hand side of (1.70) becomes

$$\frac{4\pi}{(4\pi Dt)^{3/2}}(4Dt)^{5/2} \int_0^\infty du \, u^4 e^{-u^2}. \tag{1.71}$$

The integral is $3\sqrt{\pi}/8$, so finally,

$$\langle x^2(t) \rangle = \frac{4\pi}{\pi^{3/2}}(4Dt)\frac{3\sqrt{\pi}}{8} = 6Dt. \tag{1.72}$$

Comparing with (1.59) we see that $6T\tau/m = 6D$, and since $\tau = m/\gamma$,

$$D = \frac{T}{\gamma} = \mu T. \tag{1.73}$$

This is the Einstein relation, which is one form of the fluctuation–dissipation theorem. It is an exceedingly important result, and there is nothing obvious about it. Equation (1.73) relates the diffusion constant of the particle to the particle's mobility. The former originates microscopically from equilibrium thermal fluctuations, as we have seen. The latter describes hydrodynamic friction (see eq. (1.46)), originating from a nonequilibrium situation where we apply an external force to drag the particle through the fluid. Certainly, some relation should exist between diffusion, which is caused by molecular collisions, and friction, which is also caused by molecular collisions, but it is remarkable that this relation can be expressed in the simple, general form (1.73).

1.5 Correlation Functions

Let us go back to the Langevin equation (1.39); for simplicity we consider it in 1-D, and use the notation $\gamma/m = 1/\tau$, $\Gamma/m = \Gamma_m$; then

$$\frac{dv}{dt} = -\frac{1}{\tau}v + \Gamma_m(t), \tag{1.74}$$

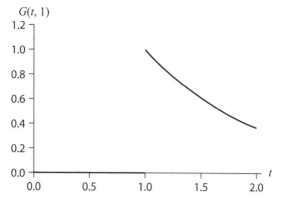

FIGURE 1.5. Graph of the function $G(t, t')$ in eq. (1.77) for a fixed value $t' = 1$.

where the random function $\Gamma_m(t)$ has dimensions of force/mass. Since $\Gamma_m(t)$ is a stochastic term, really we want to solve (1.74) for *any* function $\Gamma_m(t)$, not a specific one. We therefore use the method of Green's functions, and consider instead the equation

$$\frac{d}{dt} G(t - t') + \frac{1}{\tau} G = \delta(t - t'). \tag{1.75}$$

Once we have a solution $G(t, t')$ of (1.75), then

$$v(t) = \int_{-\infty}^{+\infty} dt' \, G(t, t') \Gamma_m(t') \tag{1.76}$$

solves (1.74) for any function Γ_m, as we can easily verify by direct substitution. For $t \neq t'$, the solution of (1.75) is $G \sim e^{-(t-t')/\tau}$, so the Green's function that solves the initial value problem (given $v(0)$, calculate $v(t)$ for $t > 0$) is

$$G = \begin{cases} e^{-(t-t')/\tau} & \text{for } t' \leq t, \\ 0 & \text{for } t' > t. \end{cases} \tag{1.77}$$

This G has a jump of size 1 at $t = t'$, so its derivative gives the δ-function in (1.75). Figure 1.5 shows $G(t - t')$ in the neighborhood of $t = t'$. We use (1.77) and (1.76), assume $t \geq 0$, and split the integral:

$$v(t) = \int_{-\infty}^{0} dt' \, G(t - t') \Gamma_m(t') + \int_{0}^{+\infty} dt' \, G(t - t') \Gamma_m(t'). \tag{1.78}$$

The first integral is

$$e^{-t/\tau} \int_{-\infty}^{0} dt' \, e^{t'/\tau} \Gamma_m(t') \equiv v(0) e^{-t/\tau}, \tag{1.79}$$

where we have called the integral in (1.79), which does not depend on t, $v(0)$. The reason will be apparent in a moment. The second integral in (1.78) is

$$\int_0^t dt'\, e^{-(t-t')/\tau} \Gamma_m(t') = e^{-t/\tau} \int_0^t dt'\, e^{t'/\tau} \Gamma_m(t'), \tag{1.80}$$

so that

$$v(t) = v(0)e^{-t/\tau} + e^{-t/\tau} \int_0^t dt'\, e^{t'/\tau} \Gamma_m(t') \tag{1.81}$$

is the required solution of (1.74) for $t \geq 0$ and any function $\Gamma_m(t)$. From (1.81) we see immediately that

$$\langle v(t) \rangle = v(0)e^{-t/\tau} \tag{1.82}$$

since $\langle \Gamma_m(t') \rangle = 0$. This says, once again, that τ is the relaxation time over which memory of the initial velocity is lost.

Now we use the solution (1.81) to calculate $\langle v^2(t) \rangle$, and require that, for $t \to \infty$, $\langle v^2 \rangle \to 3T/m$ (the equipartition result), independent of initial conditions. The result will be a relation between τ, which describes dissipation, and the "strength" of Γ_m, which describes fluctuations: one form of the fluctuation–dissipation theorem. From (1.81),

$$\langle v^2(t) \rangle = v(0)^2 e^{-2t/\tau} + 2v(0)e^{-2t/\tau} \int_0^t dt'\, e^{t'/\tau} \langle \Gamma_m(t') \rangle$$

$$+ e^{-2t/\tau} \int_0^t dt' \int_0^t dt''\, e^{(t'+t'')/\tau} \langle \Gamma_m(t') \Gamma_m(t'') \rangle. \tag{1.83}$$

The second term on the right-hand side is zero because $\langle \Gamma(t) \rangle = 0$. The third term contains the *correlation function*

$$C_m(t, t') \equiv \langle \Gamma_m(t) \Gamma_m(t') \rangle. \tag{1.84}$$

For a stationary process (time-translation invariant), the correlation function must be a function of the time difference only:

$$C_m(t, t') = C_m(t - t') \equiv C_m(s), \tag{1.85}$$

and $C_m(0) = \langle \Gamma_m^2 \rangle$ is a constant (independent of t). For the random walk problem, $C_m(s) \to 0$ for $s > \tau_0$; recall that τ_0 is the time between steps (see (1.1)), or in more physical terms, the mean free path over the thermal velocity. The reason is that $\Gamma_m(t)$ and $\Gamma_m(t+s)$ are uncorrelated for $s > \tau_0$, so $\langle \Gamma_m(t) \Gamma_m(t+s) \rangle = \langle \Gamma_m(t) \rangle \langle \Gamma_m(t+s) \rangle = 0$. The integral in (1.83), changing variables according to $t'' = t' + u$, becomes

$$\int_0^t dt' \int_{-t'}^{-t'+t} du\, e^{(2t'+u)/\tau} \langle \Gamma_m(t') \Gamma_m(t'+u) \rangle$$

$$= \int_0^t dt'\, e^{(2t')/\tau} \int_{-t'}^{-t'+t} du\, e^{u/\tau} C_m(u) \tag{1.86}$$

using (1.84). In the second integral on the right-hand side of (1.86), the integrand is nonzero only for $u \approx 0$ ($|u| < \tau_0$), and where the integrand is nonzero, $e^{u/\tau} \approx 1$ (since $\tau \gg \tau_0$). So we replace that integral with

$$\int_{-\infty}^{+\infty} du\, C_m(u) \equiv C_m \tag{1.87}$$

and obtain, for the right-hand side of (1.86),

$$\int_0^t dt'\, e^{2t'/\tau} C_m = \frac{\tau}{2}(e^{2t/\tau} - 1)C_m. \tag{1.88}$$

Finally,

$$\langle v^2(t) \rangle = v^2(0)e^{-2t/\tau} + \frac{\tau}{2}C_m(1 - e^{-2t/\tau}). \tag{1.89}$$

We require that for long times ($t \gg \tau$), $\langle v^2(t) \rangle$ converges to the equipartition value $3T/m$, and we obtain

$$\frac{\tau}{2}C_m = \frac{3T}{m} \quad \Rightarrow \quad C_m = \frac{6T}{m\tau}. \tag{1.90}$$

In terms of the *force* $\Gamma(t) = m\Gamma_m(t)$ and the corresponding correlation function $C = m^2 C_m$,

$$C = \int_{-\infty}^{+\infty} du\, C(u) = 6T\frac{m}{\tau} = 6\gamma T. \tag{1.91}$$

The relation (1.91) between the correlation function of the random force Γ and the dissipation parameter γ (or the viscous timescale τ) is another form of the fluctuation–dissipation theorem. In this form, the theorem informs the numerical simulation of thermodynamic systems through Langevin dynamics. To generate trajectories on the computer for the process

$$\frac{dv}{dt} = -\frac{1}{\tau}v + \Gamma_m(t), \tag{1.92}$$

where Γ_m represents thermal noise and τ the corresponding dissipation, one would use a stochastic, uncorrelated variable

$$\langle \Gamma_m(t)\Gamma_m(t')t \rangle = -\frac{6T}{\tau}\delta(t - t'), \tag{1.93}$$

meaning that the difference equation to be iterated on the computer is

$$\Delta v = -\frac{v}{\tau}\Delta t + \Gamma_m \Delta t, \tag{1.94}$$

where

$$\Gamma_m = \sqrt{\frac{6T}{\tau}}\, 3R \tag{1.95}$$

and R is a random number uniformly distributed in $[-1, 1]$ (therefore $\langle R^2 \rangle = 1/3$). The point is that if (1.92) is to represent a thermodynamic system,

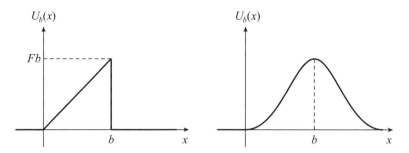

FIGURE 1.6. A linear barrier of width b and slope F (left), and a more generic barrier (right).

meaning that Γ_m represents equilibrium thermal fluctuations at a given temperature T, then Γ_m and τ in (1.94) cannot be assigned independently: they must be in the relation (1.95). Conversely, the fluctuation–dissipation theorem (1.91) can be used as a (necessary) test for determining whether fluctuations in a given thermodynamic system (the shape of a living cell, say) are thermal in origin or else driven by some other nonequilibrium process.

1.6 Barrier Crossing

One fundamental phenomenon, which appears as a consequence of thermal fluctuations, is barrier crossing. In classical mechanics (i.e., at zero temperature), a particle either has enough kinetic energy to climb over a barrier, or it doesn't. But at finite temperature, a particle always has a chance to cross a barrier, being propelled by thermal fluctuations. Under some conditions, one can obtain general expressions for the current across a barrier which do not depend on all the details of the barrier's shape. Therefore, we start with a very simple barrier in 1-D: a potential energy of the form (figure 1.6)

$$U_b(x) = \begin{cases} Fx & \text{for } 0 < x < b, \\ 0 & \text{otherwise.} \end{cases} \tag{1.96}$$

We imagine a density of Brownian particles along the line, which is uniform away from the barrier, and equal to ρ_0. Then the net current of particles across the barrier is of course zero. To obtain a steady current, we may introduce an external force field acting on the particles, described by the potential energy $U_e = -fx$. The potential energy seen by the particles is now

$$U(x) = U_b + U_e = \begin{cases} Fx - fx & \text{for } 0 < x < b, \\ -fx & \text{otherwise.} \end{cases} \tag{1.97}$$

In general, the current has the form

$$j(x) = -D\frac{\partial\rho}{\partial x} - \mu\frac{\partial U}{\partial x}\rho = -D\left(\frac{\partial\rho}{\partial x} + \frac{1}{T}\rho\frac{\partial U}{\partial x}\right),$$ (1.98)

where ρ is the concentration of particles, D is the diffusion constant, μ is the mobility, and we have used the Einstein relation $D = \mu T$ in the second equality. The first term in (1.98) is the current driven by the concentration gradients (see (1.65)), and the second is the current driven by the (average) forces on the particles (see (1.46)). Since

$$\frac{\partial}{\partial x}(\rho e^{U/T}) = e^{U/T}\left(\frac{\partial\rho}{\partial x} + \frac{\rho}{T}\frac{\partial U}{\partial x}\right),$$ (1.99)

we may rewrite (1.98) in the form

$$j(x) = -De^{-U/T}\frac{\partial}{\partial x}(\rho e^{U/T}).$$ (1.100)

In the steady state that results from applying the constant force f (see (1.97)), $j(x)$ is independent of x (since $\vec{\nabla}\cdot\vec{j} = 0$) and $\rho(x)$ is independent of x outside the interval $(0, b)$: in particular, $\rho(0) = \rho(b) = \rho_0$. Rearranging (1.100) and integrating from 0 to b gives ($j(x) = j$ being a constant)

$$\frac{j}{D}\int_0^b dx\, e^{U(x)/T} = -[\rho(x)e^{U(x)/T}]_0^b = -\rho_0(e^{-fb/T} - 1),$$ (1.101)

the last form using (1.97) and the boundary conditions on the concentration ρ. Thus, in general, for the current we have

$$j = -D\frac{[\rho(x)e^{U(x)/T}]_0^b}{\int_0^b dx\, e^{U(x)/T}}.$$ (1.102)

With the potential energy (1.97), the integral in the denominator is

$$\frac{T}{F-f}[e^{(F-f)b/T} - 1],$$ (1.103)

while the expression in the numerator is

$$\rho_0(e^{-fb/T} - 1),$$ (1.104)

giving, for the current,

$$j = \frac{D\rho_0}{T/(F-f)}\frac{1 - e^{-fb/T}}{e^{(F-f)b/T} - 1}.$$ (1.105)

For the important case of a large barrier ($Fb/T \gg 1$) and a weak external field ($f \ll F$), the fraction in (1.105) simplifies to

$$\frac{e^{fb/T} - 1}{e^{Fb/T}[1 - e^{(f-F)b/T}]} \approx e^{-Fb/T}(e^{fb/T} - 1),$$ (1.106)

and we obtain the current across the barrier as

$$j = \frac{D\rho_0}{T/F} e^{-Fb/T} \left(e^{fb/T} - 1 \right). \tag{1.107}$$

We notice that this (asymmetric) barrier works as a diode: for f large ($|fb/T| > 1$) and positive there is a relatively large current to the right, whereas for f large and negative there is only a small current to the left. The other features of (1.107) are general, not specific to this particular barrier shape. Namely, the current is proportional to the diffusion constant D, to the Arrhenius factor $e^{-Fb/T}$, where $Fb \equiv \Delta$ is the barrier height, while $T/F \equiv \lambda$ is the width of the barrier measured $\sim 1\,\mathrm{kT}$ below the maximum. In terms of these quantities, the current (which is a rate because we are 1-D) is

$$j = \frac{D\rho_0}{\lambda} e^{-\Delta/T} \left(e^{fb/T} - 1 \right). \tag{1.108}$$

A different way of obtaining a steady state current is to impose a concentration difference between the two sides of the barrier, instead of imposing an external field. Suppose we maintain the particle concentration $\rho = \rho_1$ to the left of the barrier and $\rho = \rho_2$ to the right, while $f = 0$. The expression (1.102) still applies, with $U(x)$ given by (1.96) and the new boundary conditions on ρ. We have

$$\left[\rho(x) e^{U(x)/T} \right]_0^b = \rho_2 - \rho_1, \qquad \int_0^b dx \, e^{U(x)/T} = \frac{T}{F} \left(e^{Fb/T} - 1 \right) \tag{1.109}$$

and therefore

$$j = \frac{D}{\lambda} e^{-\Delta/T} (\rho_1 - \rho_2). \tag{1.110}$$

If $\rho_1 > \rho_2$ we have a current to the right, as expected. Notice that if we drive the current with a concentration gradient, then the same barrier does *not* act as a diode.

If there is both a concentration gradient *and* an external field, we can easily find that the current is given by

$$j = \frac{D}{\lambda} e^{-\Delta/T} \left(\rho_1 e^{fb/T} - \rho_2 \right). \tag{1.111}$$

We can balance a concentration gradient with an external field; the condition of zero current gives

$$\frac{\rho_1}{\rho_2} = e^{fb/T}, \tag{1.112}$$

which is the equilibrium statistical mechanics result, fb being the work done by the external field on a particle that moves from the left to the right of the barrier.

Let us now see that these results hold also for a more general barrier shape, as in figure 1.6 (right). We consider a steady state situation with particle concentration $\rho = \rho_0$ to the left of the barrier and $\rho = 0$ to the right, with no external field. Starting from the general expression (1.102), we now have $\rho(0) = \rho_0$, $\rho(b) = 0$, and the potential energy U reflects only the barrier of figure 1.6, so $U(0) = 0$ while $U(b)$ is the barrier height. The numerator of the fraction in (1.102) is therefore equal to $-\rho_0$. In the case $U(b)/T \gg 1$ that we are considering, the integral in the denominator of (1.102) is dominated by values of x close to b; expanding $U(x)$ around $x = b$,

$$U(x) = U(b) + \frac{1}{2}U''(b)(x - b)^2 + \cdots, \tag{1.113}$$

we obtain

$$\int_0^b dx\, e^{U(x)/T} \approx e^{U(b)/T} \int_0^b dx \exp\left(\frac{U''(b)}{2T}(x - b)^2\right). \tag{1.114}$$

In the integral on the right-hand side of (1.114) the integrand is a Gaussian centered at $x = b$ (since $U''(b) < 0$), and under our assumptions we may write

$$\int_0^b dx \exp\left(\frac{U''(b)}{2T}(x - b)^2\right) = \frac{1}{2}\int_{-b}^{+b} dx \exp\left(\frac{U''(b)}{2T}x^2\right)$$

$$\approx \frac{1}{2}\int_{-\infty}^{+\infty} dx \exp\left(\frac{U''(b)}{2T}x^2\right)$$

$$= \sqrt{\frac{2T}{-U''(b)}}\frac{\sqrt{\pi}}{2}, \tag{1.115}$$

so that, ignoring numerical factors of order 1, we have for the current,

$$j = D\rho_0\sqrt{\frac{-U''(b)}{2T}}\, e^{-U(b)/T}, \tag{1.116}$$

where $U(b) \equiv \Delta$ is the barrier height, while $\sqrt{2T/-U''(b)} \equiv \lambda$ is the (half-)width of the barrier at the height $U(b) - kT$; so again we may write

$$j = \frac{D\rho_0}{\lambda}e^{-\Delta/T}, \tag{1.117}$$

which is the same as (1.110) since we have set $\rho_1 = \rho_0, \rho_2 = 0$.

If instead of using a concentration gradient, we drive the current with an external field: we set $\rho = \rho_0$ to the left and right of the barrier, and use the potential energy

$$U = U_b + U_e = \begin{cases} U_b(x) - fx & \text{for } 0 < x < c, \\ -fx & \text{otherwise,} \end{cases} \tag{1.118}$$

where $U_b(x)$ represents a generic barrier as depicted in figure 1.6, with $U_b(x) = 0$ for $x \le 0$ or $x \ge c$, $x = b$ being the position of the maximum. In the steady state, eq. (1.102) holds, but we need to specify $\rho(b)$. We assume a quasi-equilibrium situation where

$$\rho(b) \approx \rho_0 e^{-U_b(b)/T}, \tag{1.119}$$

which is justified if $fb \ll U(b)$ (weak external field). That is, we assume that applying the external field does not perturb significantly the equilibrium distribution $\rho(x)$. Then the numerator in (1.102) is

$$\rho(x)e^{U(x)/T}\Big|_0^b = \rho_0\left(e^{-fb/T} - 1\right), \tag{1.120}$$

while for the integral in the denominator, using the same quadratic approximation (1.113) for $U_b(x)$, we have

$$\int_0^b dx\, e^{U(x)/T} \approx e^{U_b(b)/T} \int_0^b dx\, e^{\left[\frac{1}{2}U_b''(b)(x-b)^2 - fx\right]/T}. \tag{1.121}$$

Consider the exponent in the integrand of (1.121): with the shorthand $U_b''(b)/2 \equiv a$ and completing the square,

$$a(x-b)^2 - fx = a\left[x - \left(b + \frac{f}{2a}\right)\right]^2 - a\left[\left(\frac{f}{2a}\right)^2 + \frac{bf}{a}\right]. \tag{1.122}$$

Under our conditions $(fb \ll U(b),\ |a|b^2 \sim U(b))$, we have $f/|a| \ll b$, so neglecting f/a compared to b in (1.122) we may write $a(x-b)^2 - fx \approx a(x-b)^2 - fb$ in (1.121) and obtain

$$\int_0^b dx\, e^{U(x)/T} \approx e^{U_b(b)/T} e^{-fb/T} \int_0^b dx\, e^{a(x-b)^2/T}. \tag{1.123}$$

The integral on the right-hand side is given in (1.115); again ignoring numerical factors of order 1 we finally obtain for the current,

$$j = D\rho_0 \sqrt{\frac{U_b''(b)}{2T}}\, e^{-[U_b(b)-fb]/T}\left(1 - e^{-fb/T}\right). \tag{1.124}$$

Written in terms of the barrier height Δ and width λ (remember, though, that λ is a temperature-dependent parameter),

$$j = \frac{D\rho_0}{\lambda}\, e^{-(\Delta-fb)/T}(1 - e^{-fb/T}). \tag{1.125}$$

From the way we have set up the problem, this formula is valid for $f \geq 0$, that is, for $j \geq 0$. The parameter b describes the ascending part of the barrier; beyond a barrier width λ from the top, the descending part of the barrier does not matter because the "extra" particles that enter it (those beyond the equilibrium distribution for zero current) are swept away by the field. Equation (1.125) is (for $f \geq 0$) the same as (1.108). If we compare (1.125) and (1.111), we see that, for $fb/T > 1$, the effect of the field f is twofold: it lowers the barrier by an amount fb (this is a geometric effect), and it sweeps away the particles injected from the left into the right-hand side of the barrier, so that, effectively, we can set $\rho_2 = 0$ in (1.111), and $\rho_1 = \rho_0$, in order to calculate the current.

Finally, we consider the problem of the rate of escape of a particle from a potential well (figure 1.7).

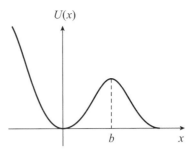

$U(x)$

b x

FIGURE 1.7. The potential energy for the problem of escape of a particle from a potential well.

This situation describes, in a more or less idealized fashion, many systems in physics and chemistry. Actually we have the answer already, namely (1.116), if for ρ_0 we put in a value that describes the concentration corresponding to one particle in the potential well: $\rho_0 \rightarrow 1/(\text{size of region explored by particle}) \sim 1/\langle x^2 \rangle^{1/2}$, with $\langle x^2 \rangle^{1/2}$ being the rms position fluctuations of the particle in the well. We calculate $\langle x^2 \rangle$ from equilibrium statistical mechanics, assuming that the particle is essentially in equilibrium in the well before it escapes (i.e., the barrier is high enough that the escape rate is small compared to the viscous relaxation time τ, and very small compared to the collision time τ_0 of the Brownian motion). The probability that the particle is at position x is

$$p(x) = \frac{1}{Z} e^{-U(x)/T}, \qquad Z = \int_{-\infty}^{+\infty} dx\, e^{-U(x)/T}, \qquad (1.126)$$

while

$$\langle x^2 \rangle = \int_{-\infty}^{+\infty} dx\, x^2 p(x). \qquad (1.127)$$

Using the quadratic approximation around $x = 0$,

$$U(x) \approx \frac{1}{2} U''(0) x^2 + \cdots, \qquad (1.128)$$

and carrying out the elementary integrals (this amounts to calculating the rms fluctuations of a spring) we find, in accordance with equipartition,

$$\langle x^2 \rangle = \frac{T}{U''(0)}, \qquad (1.129)$$

and therefore we obtain from (1.116), up to numerical factors of order 1,

$$j = D \sqrt{\frac{U''(0)}{T}} \sqrt{\frac{-U''(b)}{2T}} e^{-U(b)/T}. \qquad (1.130)$$

This is the Kramers formula for the rate of escape over a potential barrier. If we define, like before, the (temperature-dependent) half-width of the barrier at the maximum ($\lambda_b = \sqrt{2T/|U''(b)|}$) and the corresponding width of the well at the minimum ($\lambda_0 = 2\sqrt{2T/U''(0)}$), the barrier height being $U(b) \equiv \Delta$, we can write (1.130) as

$$j = 2\sqrt{2}\,\frac{D}{\lambda_0\lambda_b}e^{-\Delta/T}. \qquad (1.131)$$

Alternatively, using the Einstein relation $D = \mu T$ we can write (1.130) as

$$j = \frac{1}{\sqrt{2}}\mu\sqrt{-U''(0)U''(b)}\,e^{-U(b)/T}. \qquad (1.132)$$

Either way, the rate of escape is inversely proportional to the viscosity η in which the particle moves, since $D, \mu \propto 1/\eta$.

Throughout this section we have considered the problem of a particle *diffusing* across a barrier; the mean free path is small compared to the barrier width. Then the rate of barrier crossing is inversely proportional to the viscosity η. However, this dependence cannot hold for $\eta \to 0$. For η sufficiently small, the rate of barrier crossing is proportional to the viscosity instead.

1.7 What is Equilibrium?

The concept of equilibrium in statistical mechanics is, to put it mildly, not straightforward. It is indissolubly tied to the idea of entropy, and the second law of thermodynamics, so that the three concepts can hardly be clarified separately. As we know, for a closed system the equilibrium state maximizes the entropy S. From this principle, the idea of equilibrium, and a microscopic definition for the entropy, one can more or less derive the basic results of statistical mechanics, such as the equilibrium properties of an open system in contact with a thermostat and so on. The second law says that, for a closed system,

$$\frac{dS}{dt} \geq 0, \qquad (1.133)$$

so if we "start" the closed system out of equilibrium (conceptually, by removing a constraint), it will typically evolve in a way to increase the entropy (however, (1.133) does not represent a guarantee that equilibrium will be reached). Now, the important thing is that, as soon as we bring in entropy, we must bring in time, as we see from (1.133) and from the very words we use to describe the significance of the entropy: if we "start" the system this way or that way, meaning that we are now discussing time. The physical concepts of entropy and equilibrium presume a certain observation time (the mathematical concepts can be discussed "instantaneously" in terms of ensembles, the two frameworks being connected by the ergodic hypothesis). To proceed, we need a microscopic definition of the entropy; take, for simplicity, the case of discrete states (e.g., a system of spins)

$$S = -\sum_i p_i \ln p_i, \tag{1.134}$$

where p_i is the probability that the system is in state i. Now consider the two-states system at temperature T. The system is open, as it is in contact with a thermostat. The two energy levels are 0 and $\varepsilon > 0$, and we call $p(0)$, $p(\varepsilon)$ the probabilities that the system is in the two respective states. We know that the *equilibrium state* minimizes the free energy F:

$$F = E - TS = \varepsilon p(\varepsilon) - T\big[p(0)\ln p(0) + p(\varepsilon)\ln p(\varepsilon)\big]. \tag{1.135}$$

Minimizing (1.135) with respect to $p(0)$, $p(\varepsilon)$ with the constraint $p(0) + p(\varepsilon) = 1$, we find of course the Boltzmann distribution:

$$p(\varepsilon) = \frac{e^{-\varepsilon/T}}{1 + e^{-\varepsilon/T}}, \qquad p(0) = \frac{1}{1 + e^{-\varepsilon/T}}. \tag{1.136}$$

The point is that microscopically, equilibrium means a particular probability distribution according to which the system visits the different states. For the system in contact with the thermostat, it is the Boltzmann distribution (1.136); for the closed system, it is the uniform distribution (all states equiprobable); etc. We see that we have to observe the system for some time in order to determine whether it is in equilibrium. Equilibrium is not a property of the microscopic states of a system, it is a property of the probability distribution. At the risk of belaboring the point, consider a gas in a box. The question, "Is the state with all molecules in the left half of the box an equilibrium state?" is meaningless; the question with respect to equilibrium is, "How often is this state visited by the system?" For this reason, entropy must be defined in terms of probabilities if it is to be a measure of how close or far we are from equilibrium. Therefore entropy, unlike energy, is not an instantaneous quantity (we are reasoning within classical physics here). And therefore, now we have the question of the observation time.

Here is what Richard Feynman writes about equilibrium and observation time, in the introduction to his lectures on statistical mechanics: "If a system is very weakly coupled to a heat bath at a given 'temperature,' if the coupling is indefinite or not known precisely, if the coupling has been on for a long time, and if all the 'fast' things have happened and all the 'slow' things not, the system is said to be in *thermal equilibrium*." Because of this connection between equilibrium and observation time, the problem of Brownian motion stands at the interface between equilibrium and nonequilibrium statistical mechanics. The trajectory $\vec{x}(t)$ of the Brownian particle evolves in time; for example, if we start observing at $t = 0$, the second moment $\langle |\vec{x}(t) - \vec{x}(0)|^2 \rangle$ increases with time, and the probability distribution $p(\vec{x})$, which is centered at $\vec{x}(0)$, keeps evolving, as we saw. If we have the particle in a box, eventually the trajectory will fill the box and the probability distribution $p(\vec{x})$ becomes uniform, provided that we observe on these long timescales. On the other hand, if we look at the momentum of the particle, and start observing at $t = 0$, we find that the corresponding probability distribution becomes time independent over much shorter timescales, of order a few molecular collision times, which are measured in picoseconds. For the problem of Brownian motion as it is normally posed, observation times are long compared to molecular collision times, but short compared to the diffusion time of the particle across the box. Therefore the momentum of the particle is "in equilibrium"; for example, the second moment has its time-independent equipartition value $\langle |\vec{p}|^2 \rangle = 3T/m$, but the position of the particle undergoes a nonequilibrium process, for example, the second moment evolves in time according to $\langle |\vec{x}(t) - \vec{x}(0)|^2 \rangle = 6DT$ (see (1.72)). In this same sense, we will see in chapter 4 that for an enzyme undergoing its catalytic cycle, the momentum of the atoms comprising the enzyme can be considered "in equilibrium," but the relative position of these atoms undergoes a nonequilibrium process, as the enzyme deforms.

2

Statics of DNA Deformations

2.1 Introduction

DNA is a deformable molecule. The term "deformable" already implies phenomena rooted in the collective behavior of many atoms, and a description based on concepts of continuum and statistical mechanics. Long DNA molecules (with a contour length of many persistence lengths) are an excellent experimental system to study the equilibrium conformations and dynamics of long, flexible molecules: the traditional focus of polymer physics. Among the many reasons is that long DNA is a heteropolymer that can be made exactly to order, monodispersed, by relatively simple, well-established molecular biology methods. Taking advantage of sequence complementarity, it can be labeled at specific sites. Short DNA molecules, on the other hand, are a model experimental system for the study of the molecular deformability we are concerned with in this book. A 30 base pairs (bp) double-stranded (ds) DNA molecule is, roughly, a compact cylinder 10 nm long and 2 nm wide, consisting of 3 turns of the double helix and about 2×10^3 atoms. It is a "soft" nanoparticle, in the sense that it can undergo very large but reversible deformations, such as partially melting into two single strands, bending way past the linear elasticity regime, and twisting. At the same time, the unperturbed, average structure of this nanoparticle is as well defined as that of a crystal, each copy of the same particle structurally identical in an ensemble-averaged sense, as far as we know. We stress *ensemble-averaged* structure because it differs from any instantaneous structure owing to generally large thermal fluctuations. This remarkable deformability of DNA nanorods stems from the fact that the atoms in this structure are bound together by interactions on two very different energy scales. The chemical structure of the single strand is maintained by covalent bonds; with characteristic bond energies

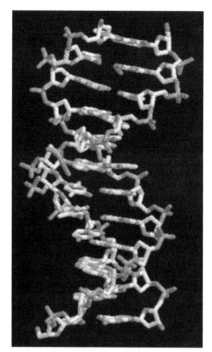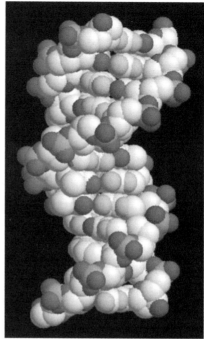

FIGURE 2.1. Structure of a DNA 12mer obtained by X-ray diffraction (PDB: 1BNA). The helix is slightly bent, owing to packing in the crystal. The phosphate groups forming the sugar–phosphate backbone of the strands appear relatively darker in this grayscale representation. The same structure is shown in two different representations ("sticks" and "all atom"), using the molecular visualization program RasMol.

of several eV (electronvolts), these are fixed and typically immutable with respect to the deformation processes we consider. In contrast, the structure of the double helix, that is, the overall structure of the nanorod, is maintained by a large set of much weaker bonds, *orthogonal* to the polymer backbone. These are mainly hydrogen bonds, dipole–dipole interactions ("Van der Waals forces"), and ionic interactions. For each of these, the characteristic bond energy is $1\,k_B T \approx 25\,\mathrm{meV}$ (in this book we use the notation $k_B T$ as an energy unit, corresponding to the Boltzmann constant times 300 K, i.e., room temperature). This is true for ionic bonds also, owing to the large dielectric constant of water ($\varepsilon_{water} \approx 80$ at low frequency). For the DNA nanorod, the relevant weak bonds can be classified as follows:

- base-pairing interactions: hydrogen bonds between Watson–Crick base pairs; 2 hydrogen bonds for A–T, 3 for G–C
- base-stacking interactions: induced dipole interactions between adjacent bases on the same strand, which are stacked like the rungs of a ladder in the double helix

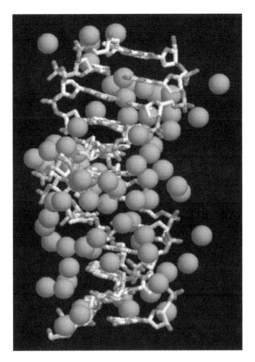

FIGURE 2.2. The same DNA structure as figure 2.1,
showing the water molecules (only the oxygen atoms)
"bound" to the structure in the crystal.

- electrostatic repulsion between the negatively charged (ionized)
  phosphate groups of the sugar–phosphate backbone (destabilizing)

Both base pairing and base stacking are stabilizing (attractive) interactions
and they are of the same order.

The nanorod is therefore held together by of the order of 100 weak bonds
($30\,\mathrm{bp} \times 2.5$ hydrogen bonds/bp = 75 hydrogen bonds of base pairing; 29
stacking bonds $\times$ 2 strands = 58 base-stacking bonds). It is the many possible
rearrangements of these weak bonds, both in geometry and strength, that
confers a sort of "reversible plasticity" to the nanoparticle. Figure 2.1 shows
the structure of a DNA 12mer obtained by X-ray diffraction (by the Dickerson
group, in 1980).

Water has a central role in defining the above intramolecular interactions
(figure 2.2), and indeed the structure, thermodynamic stability, and dynamics
of the nanorod (and other biomolecules). For example, the meaning of
the statement "the strength of the A–T hydrogen bonding (Watson–Crick
base pairing) is $2\,k_\mathrm{B}T$" is that the free energy difference between the A–T
hydrogen bonds in the ds structure and the hydrogen bonds that A and T

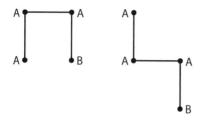

FIGURE 2.3. A polymer of 3 links on a square lattice. Left: the "folded" conformation (A and B are bound). Right: one of 4 different "unfolded" conformations.

would form with water molecules in the single-stranded (ss) DNA is $2\,k_B T$. The strength of an internal hydrogen bond depends on how accessible it is to water. The free energy difference between two different structural states depends on the water accessibility of various groups for these two states. For this reason, intramolecular interactions are effectively temperature dependent, water accessibility being different at different temperatures. To put it differently, surface energies are as important as bulk energies, the "surface" being the water–structure interface. Different conformations of the molecule have different surfaces.

These are not small effects: for example, a salt (NaCl) crystal in air melts at approximately 800 C, while in water it melts at room temperature. DNA and proteins do not melt at room temperature because the atoms are bound together in a covalently linked polymer chain. The basic design is the same for proteins and DNA: the polymer chain folds on itself (or, in the case of DNA, twists around another chain) creating an interior volume that excludes the water. Without the polymer backbone, no such structure can be stable against a zero concentration of constituents in solution, just like micelles do not form below a critical bulk concentration of components.

The role of the polymer backbone is easily appreciated through simple counting. Consider a "polymer" of 3 links (4 "monomers") on a square lattice of $N \times N$ sites (figure 2.3). Monomer B is different from the A monomers, and the A–B interaction energy is $-\varepsilon$ ($\varepsilon > 0$), when A and B are nearest neighbors. There are 1 "folded" conformation and 4 different unfolded ones (figure 2.3). The number of folded states on the $N \times N$ lattice is (disregarding boundary effects) $\Gamma_f = 4 \times N^2$. The factor 4 counts rotations and N^2 counts translations of the molecule.

The number of unfolded states is $\Gamma_u = 4 \times 4 \times N^2$. Correspondingly, the free energies are

$$F_f = -\varepsilon - T\ln(4N^2), \qquad F_u = -T\ln(4 \times 4N^2), \tag{2.1}$$

$$\Delta F = F_u - F_f = \varepsilon - 2T\ln(2). \tag{2.2}$$

For $\Delta F > 0$, the folded conformation is stable, and this needs a binding energy ε of only a few $k_B T$ at room temperature ($\varepsilon > (2\ln 2)\,k_B T$ according to (2.2)).

Compare to the case of two unattached particles A and B: the number of dimer states is $\Gamma_D = 4 \times N^2$, the number of monomer states $\Gamma_M = N^2(N^2 - 4)$

and for the free energies we now have

$$F_D = -\varepsilon - T\ln(4N^2),$$

$$F_M = -T\ln[N^2(N^2 - 4)] = -T\ln(N^2) - T\ln(N^2 - 4), \qquad (2.3)$$

$$\Delta F = F_M - F_D = \varepsilon - 2T\ln N + 2T\ln 2 \approx \varepsilon - 2T\ln N. \qquad (2.4)$$

For the dimer to be stable at room temperature now requires a huge binding energy $\varepsilon > (2\ln N)k_B T$.

In terms of the fraction of folded molecules f,

$$f = \frac{Z_f}{Z_f + Z_u} = \frac{1}{1 + Z_u/Z_f} = \frac{1}{1 + e^{-\Delta F/T}}, \qquad (2.5)$$

where Z_f, Z_u are the corresponding partition sums. For example, $f = \frac{1}{2}$ for $\Delta F = 0$, which means $\varepsilon = (2\ln 2)T$ for the polymer but $\varepsilon = (2\ln N)T$ for the independent particles. The $\ln N$ term expresses the concentration dependence (the dilution term) of the chemical potential, and is ultimately the reason why life is made of polymers!

2.2 DNA Melting

At sufficiently high temperature (perhaps around $70\,C$ for a typical DNA 30mer), the double helix dissociates into its two separate strands: the nanorod "melts." The melting temperature depends on the base sequence (since a G–C pairing is about 1.5 times stronger than an A–T pairing), DNA concentration (since one duplex dissociates into two separate strands), ionic strength of the solution (since each strand is charged), and pH (which controls the dissociation state of various groups and thus modulates the charge). We discuss this transition for DNA oligonucleotides by way of introducing the internal degrees of freedom relevant for the structure of the DNA nanorod.

Two-states model: The simplest description for the melting transition is the dissociation of two particles A and B (the two strands) with no internal structure:

$$AB \underset{k_a}{\overset{k_d}{\rightleftharpoons}} A + B. \qquad (2.6)$$

Given dissociation and association rates k_d and k_a, we write a rate equation:

$$\frac{d}{dt}[AB] = -k_d[AB] + k_a[A][B]. \qquad (2.7)$$

Let us agree that [ ] stands for concentration in molar (moles/liter: $1\,M = 1\,mole/L$), and note that k_d and k_a have different dimensions (k_d has units of s^{-1}, k_a of $L\,s^{-1}$). At equilibrium, $(d/dt)[AB] = 0$, which yields the law of mass action:

$$\frac{[A]_{eq}[B]_{eq}}{[AB]_{eq}} = \frac{k_d}{k_a} = K_d(T). \tag{2.8}$$

The subscript "eq" means that these are equilibrium concentrations, and K_d is the dissociation constant, which depends on temperature (and pressure of course, though we omit it). The fraction of melted (dissociated) nanorods is

$$p = \frac{[A]}{[A] + [AB]} = \frac{[B]}{[AB] + [B]} \tag{2.9}$$

(the number of A strands in the ss state divided by the total number of A strands, or similarly for B); we assume stoichiometric amounts of the two strands, that is, $[A] = [B]$ at all times. From this model we can see why the melting transition depends on DNA concentration. The total DNA concentration (as single strands) is

$$C = [A] + [B] + 2[AB] = \text{const.} \tag{2.10}$$

We write $[A]_{eq}$ in terms of C using (2.10) and (2.8) (and $[A] = [B]$):

$$[A]_{eq}^2 + K_d[A]_{eq} - \frac{K_d}{2}C = 0, \tag{2.11}$$

with solution

$$[A]_{eq} = [B]_{eq} = \frac{K_d}{2}\left[-1 + \sqrt{1 + \frac{2C}{K_d}}\right]. \tag{2.12}$$

At equilibrium, the fraction of melted nanorods is therefore (from (2.9), (2.8), (2.12))

$$p = \frac{1}{1 + [A]_{eq}/K_d} = \frac{1}{1 + \frac{1}{2}\left[-1 + \sqrt{1 + \frac{2}{K_d}C}\right]}, \tag{2.13}$$

which gives, for a fixed temperature, the concentration dependence $p = p(C)$ (figure 2.4). The dissociation constant $K_d(T)$ of this two-states model can thus be determined in principle by measuring p for different values of C, and using (2.13), though in practice it is easier to relax the stoichiometric condition $[A] = [B]$ and titrate one strand into a solution of the other.

The *melting point* is defined as $p = \frac{1}{2}$, and from (2.13) we obtain

$$K_d(T_m) = \frac{C}{4}. \tag{2.14}$$

If we know the temperature dependence of K_d, eq. (2.14) tells us the concentration dependence of the melting temperature T_m. The main temperature dependence of K_d can be obtained by writing down the chemical potentials

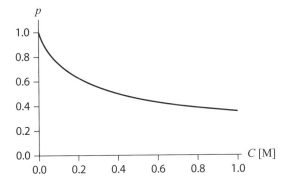

FIGURE 2.4. Plot of eq. (2.13) for $K_d = 0.1$ M; the concentration C is in M.

of the various species:

$$\begin{cases} \mu_A = \mu_A^0 + T \ln X_A, \\ \mu_B = \mu_B^0 + T \ln X_B, \\ \mu_{AB} = \mu_{AB}^0 + T \ln X_{AB}, \end{cases} \qquad (2.15)$$

where X_A is the mole fraction of species A, and μ_A^0 is a constant (meaning a concentration-independent term) which we explain below. Equation (2.15) exhibits the concentration dependence of the chemical potential, written for a dilute solution, which is equivalent to the ideal gas expression. Thus (2.15) is valid only for mole fractions $X \ll 1$, and the often repeated statement that μ^0 in (2.15) is the "chemical potential of the pure component" (since formally $\mu = \mu^0$ for $X = 1$) is not to be taken literally, because (2.15) is not valid anywhere close to $X = 1$. Since we consider dilute solutions in water, the relation between mole fraction and concentration is

$$X_A = \frac{N_A}{N_{tot}} = \frac{N_A}{N_A + N_B + N_{AB} + N_w} \approx \frac{N_A}{N_w} = \frac{[A]}{[W]}, \qquad (2.16)$$

that is,

$$X_A = \frac{[A]}{C_w}, \qquad (2.17)$$

where N is the number of particles (in the given volume) and w stands for water; $N_A, N_B, N_{AB} \ll N_w$, and the concentration of water in water is $C_w = 55$ M.

Therefore we can write $\mu_A = \mu_A^0 + T \ln[A] - T \ln C_w$, and evaluating this relation for $[A] = 1$ M and at room temperature, where $T = 25$ meV, since $\ln 55 \approx 4$ we find $\mu_A + 100$ meV $= \mu_A^0$. So we can say, for example, that μ_A^0 represents the reference chemical potential of A at a concentration of 1 M, plus 100 meV.

Coming back to (2.15), the equilibrium condition for the "reaction" (2.6) is

$$\mu_{AB} - (\mu_A + \mu_B) = 0, \tag{2.18}$$

which yields, in equilibrium,

$$\ln \frac{X_A X_B}{X_{AB}} = \frac{1}{T} \left(\mu_{AB}^0 - \mu_A^0 - \mu_B^0 \right), \tag{2.19}$$

or, using (2.17),

$$\ln \frac{[A]_{eq}[B]_{eq}}{[AB]_{eq}} = -\frac{\Delta\mu^0}{T} + \ln C_w, \tag{2.20}$$

that is, the law of mass action (2.8) once again, where

$$K_d = C_w e^{-\Delta\mu^0/T} \tag{2.21}$$

and $\Delta\mu^0 = \mu_A^0 + \mu_B^0 - \mu_{AB}^0$. To obtain the approximate temperature dependence of K_d, we write the chemical potential $\Delta\mu^0$ in terms of enthalpy and entropy,

$$\Delta\mu^0 = \Delta H^0 - T\Delta S^0 \tag{2.22}$$

and assume that ΔH^0, ΔS_0 are roughly temperature independent (this is generally false for processes in water, and this is the weakest and most uncontrolled approximation in the whole argument leading to the Van 't Hoff relation below). Then we have

$$\ln \frac{K_d(T)}{C_w} = -\frac{\Delta H^0}{T} + \Delta S^0. \tag{2.23}$$

From (2.14) and (2.23) we can now find how the melting temperature T_m depends on DNA concentration C:

$$T_m = -\frac{\Delta H^0}{\Delta S^0 - \ln\left(\frac{C}{4C_w}\right)}. \tag{2.24}$$

By measuring T_m for different concentrations C, or by measuring the temperature dependence of K_d, one can thus determine the enthalpy and entropy parameters ΔH^0 and ΔS^0, using (2.24) and (2.23), respectively.

In summary, the concentration dependence of the melting temperature is given by an additive term to the dissociation entropy ΔS^0, an additive term that is logarithmic in the concentration.

If we are interested in the temperature dependence rather than the concentration dependence of the transition, we start from a simpler and more familiar formulation of the two-states model in statistical mechanics. The two states in question are associated (AB), to which we assign the reference energy and entropy $\varepsilon_{ass} = 0$, $s_{ass} = 0$, and dissociated (A + B), with energy

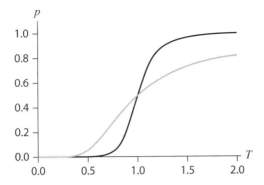

FIGURE 2.5. Plot of eq. (2.26) for two cases:
$\varepsilon_0 = s_0 = 10$ (steeper curve) and $\varepsilon_0 = s_0 = 3$. In both
cases, $T_m = \varepsilon_0/s_0 = 1$.

$\varepsilon_{\text{diss}} = \varepsilon_0 > 0$ and entropy $s_{\text{diss}} = s_0 > 0$. The partition sum is

$$Z = 1 + e^{s_0} e^{-\varepsilon_0/T}, \tag{2.25}$$

e^{s_0} being the degeneracy of the dissociated state relative to the associated one. The fraction p of dissociated molecules is equal to the probability of the dissociated state (since we consider only two states), so

$$p = \frac{1}{Z} e^{s_0} e^{-\varepsilon_0/T} = \frac{1}{1 + e^{(\varepsilon_0 - Ts_0)/T}} \tag{2.26}$$

with the melting temperature given by

$$T_{\text{m}} = \frac{\varepsilon_0}{s_0}. \tag{2.27}$$

The two-states expression (2.26) is often used to analyze melting curves of oligonucleotides (assuming ε_0 and s_0 to be temperature-independent parameters). Given T_{m}, the ratio ε_0/s_0 is fixed and the steepness of the melting curve in model (2.26) is determined by the magnitude of ε_0 and s_0 (figure 2.5). The slope at the midpoint of the transition is

$$\left.\frac{dp}{dT}\right|_{T=T_{\text{m}}} = \frac{s_0}{4T_{\text{m}}}, \tag{2.28}$$

proportional to s_0 (and thus ε_0) for fixed T_{m}. In terms of the single base pairing energy ε and entropy s, the total binding energy ε_0 and entropy s_0 are at least approximately additive, that is, for an oligonucleotide of N base pairs, $\varepsilon_0 = N\varepsilon$, $s_0 = Ns$. Therefore the two-states model (2.26) says that the longer the oligo, the steeper the melting curve. In fact,

$$p = \frac{1}{1 + e^{N(\varepsilon_0 - Ts_0)/T}} \tag{2.29}$$

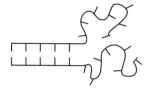

FIGURE 2.6. A partially melted DNA molecule according to the zipper model. Five base pairs are closed, and seven are open. The single-stranded parts are in a random coil conformation.

becomes a step function around $T_{\mathrm{m}} = \varepsilon/s$ for $N \to \infty$. This model exhibits by construction "perfect cooperativity": the base-pairing bonds break all together or not at all.

At the other extreme, a totally non-cooperative melting transition would be one in which each base-pairing bond is closed or open independently of all the others, that is, a situation of N independent bonds. Each open bond has an energy cost ε and entropy gain s. This is the same problem as N independent spin $\frac{1}{2}$ particles in a magnetic field. The partition sum for the system of size N is

$$Z_N = Z_1^N, \qquad Z_1 = 1 + e^{(s - \varepsilon/T)}. \qquad (2.30)$$

If n is the number of open bp, the energy is

$$E = \langle n \rangle \varepsilon = T^2 \frac{\partial \ln Z_N}{\partial T} \qquad (2.31)$$

$$\Rightarrow \quad \langle n \rangle = N \frac{1}{e^{(\varepsilon/T - s)} + 1}. \qquad (2.32)$$

Comparing $\langle n \rangle / N$ given by (2.32) with (2.29), we see that the totally cooperative transition is a factor N steeper than the totally non-cooperative transition.

Reality is in between. The nanorod does not melt in an "all or none" fashion: for temperatures near T_{m} there is, in equilibrium, a finite fraction of oligomers that are partially melted (i.e., with some broken and some intact base pairs). On the other hand, the melting curve of the nanorod is way steeper than (2.32). To do better, we must give more attention to the internal degrees of freedom of the nanorod. The steepness of the melting curves of oligonucleotides signals a degree of "cooperativity" of the transition: the fact that if one bp is open (base-pairing bond broken), it is energetically easier for the adjacent bp to open. The simplest model incorporating this effect is the zipper model.

2.3 Zipper Model

We consider a homogeneous sequence and impose the constraint that base pairs can open only in a contiguous row starting from one end (like a zipper: figure 2.6). Therefore we can classify the microscopic states of the system by the number of open bp, n. Opening each bp has an energy cost ε and an entropy gain s. The energy ε reflects the base-pairing energy and should be of order 2–3 kT. A fixed entropy gain per open bp, s, reflects the following physics: double-stranded DNA is semirigid on the length scales of the oligomer, while ss DNA is very flexible. We think of the ds part of a partially melted molecule as occupying one conformational state, and of each strand of the melted part as a random coil, occupying a number of conformational states corresponding to a random walk of n steps. With g states per step, the number of states for the random walk is g^n, that is, an entropy $n \ln g$ or an entropy per step $s = \ln g$.

The partition sum for the zipper model is

$$Z = \sum_{n=0}^{N} g^n e^{-n\varepsilon/T} = \sum_{n=0}^{N} e^{n(s-\varepsilon/T)} \tag{2.33}$$

(for simplicity, we are omitting at present an extra entropy contribution in the $n = N$ term which comes from strand dissociation; that term carries the concentration dependence of the transition, as we saw before, which is therefore not present in (2.33)). The geometric series can be summed in closed form:

$$Z = \frac{1 - e^{(N+1)(s-\varepsilon/T)}}{1 - e^{(s-\varepsilon/T)}}. \tag{2.34}$$

The probability that n bp are open is

$$p(n) = \frac{1}{Z} e^{n(s-\varepsilon/T)} \tag{2.35}$$

and the average number of open bp is

$$\langle n \rangle = \frac{1}{Z} \sum_{n=0}^{N} n e^{n(s-\varepsilon/T)} = \frac{\partial}{\partial s} \ln Z. \tag{2.36}$$

The case of an infinitely long molecule is simplest:

$$\text{for } N \to \infty, \quad Z \to \frac{1}{1 - e^{(s-\varepsilon/T)}}. \tag{2.37}$$

Thus for $T \leq T_c = \varepsilon/s$, and taking the derivative according to (2.36), we find

$$\langle n \rangle = \frac{1}{e^{\varepsilon/T - s} - 1} \quad (T \leq T_c), \tag{2.38}$$

which diverges at T_c (figure 2.7). So in the thermodynamic limit $N \to \infty$, the zipper model has a phase transition at the critical temperature T_c, signalled by the divergence of the partition sum (2.37). Note that the Landau argument for nonexistence of phase transitions in one-dimensional systems with short-range interactions does not apply here: the zipper model has effectively infinite-range interactions.

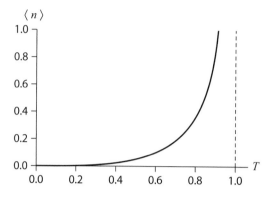

FIGURE 2.7. Plot of eq. (2.38) for $\varepsilon = s = 1$ ($T_c = 1$).

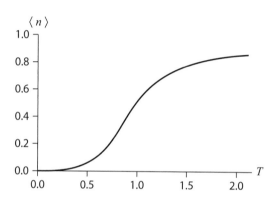

FIGURE 2.8. Plot of eq. (2.39) for $\varepsilon = s = 1$ and $N = 10$ ($T_m = 1$).

However, we are interested in the finite system, for which we find, using (2.34) and (2.36),

$$\langle n \rangle = \frac{N+1}{1 - e^{-(N+1)(s-\varepsilon/T)}} - \frac{1}{1 - e^{-(s-\varepsilon/T)}}. \qquad (2.39)$$

This function is plotted in figure 2.8. Now $T_m = \varepsilon/s$ is not a critical point, but it is the melting temperature in the sense that

$$\langle n \rangle (T = T_m) = \frac{N}{2}, \qquad (2.40)$$

which we can verify by taking the limit $(s - \varepsilon/T) \to 0$ in (2.39).

2.4 Experimental Melting Curves

Optical absorption by the bases in DNA around 260 nm provides a convenient method to obtain melting profiles. The dipole transition responsible for

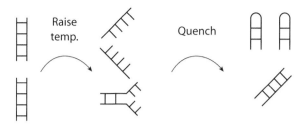

FIGURE 2.9. Principle of the quenching method: partially melted molecules re-form duplexes after the quench, while dissociated molecules form hairpins. Adapted from Zeng, Montrichok, and Zocchi (2004).

absorption is somewhat quenched when the bases are stacked; as a result, base unstacking results in increased UV absorption. Base unpairing also results in a similar increase in the absorption coefficient. Melting curves can be easily obtained using μM concentrations of DNA oligomers and a commercial spectrophotometer. Thus the UV absorption signal reports on the fraction of bases in the sample that are unstacked and/or unpaired.

With finite-size molecules, unpaired bases in the sample may belong either to molecules that are completely dissociated into two separate strands, or to molecules in intermediate, "bubble state" conformations where part of the molecule is in the ds form and part in the ss form. The presence of such intermediate states can be quantified by measuring, besides the fraction of open base pairs f, the fraction of completely dissociated molecules, p. The following "quenching method" provides the dissociation curve $p(T)$. Sequences are chosen to be partially self-complementary, so individual strands can form hairpins. However, the ground state, accessed by careful annealing, is still the duplex structure. Starting from the duplex at "low" temperature (well below the melting transition, say 0 C), the sample is brought to an intermediate temperature T (within the transition region), equilibrated, then brought back quickly ("quenched") to 0 C. Molecules that were dissociated at temperature T form hairpins, while molecules that were in a bubble state form the ground state duplex after the quench (figure 2.9). The relative amount of hairpins and duplexes is measured by gel electrophoresis (hairpins are shorter and move faster through the gel); see figure 2.10. In summary, the quenching method reports on the fraction of dissociated molecules $p(T)$.

A glance at an experimental melting curve of oligomers obtained by UV absorption (figure 2.11) shows that the process is more complex than the simple models we have discussed so far. It is obvious from figure 2.11(A1) that there are *two* melting processes, and indeed, there are really two different structures that "melt." One melting process is unpairing of the base pairs (breaking of the bonds between complementary bases on opposite

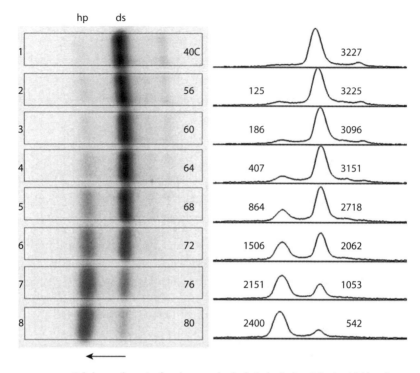

FIGURE 2.10. Gel electrophoresis after the quench: the hairpin (hp) and duplex (ds) bands are indicated. The temperature T to which the sample was brought before the quench is indicated for each lane. As T is increased, the duplex band disappears and the hairpin band appears, signifying an increase in the dissociated molecules $p(T)$. On the right is a plot of the intensity profile of the lanes, averaged across the lane. Adapted from Zeng, Montrichok, and Zocchi (2004).

strands); the other melting process, which happens at somewhat higher but partially overlapping temperatures, is unstacking of the adjacent bases on the same strand. The UV absorption curve figure 2.11(A1) reflects both: the region $45 < T < 55\,\mathrm{C}$ corresponds to melting of the ds structure, as we see from the strand dissociation curve figure 2.11(A2), which shows that strand dissociation is complete ($p = 1$) at $T = T_c \approx 55\,\mathrm{C}$; the region $T > 65\,\mathrm{C}$ corresponds to unstacking of the bases in the single strands. The midpoint of this unstacking transition is above $100\,\mathrm{C}$. We denote by T_c the temperature at which there is (essentially) 100% strand dissociation, since for the infinite molecule ($N \to \infty$) it becomes the critical point.

The melting curve (A1) is a little extreme in that the unstacking transition is more prominent than the unpairing transition; we chose this deliberately to make the point that stacking degrees of freedom are as important as pairing degrees of freedom in the structural transitions of DNA. The reason for the

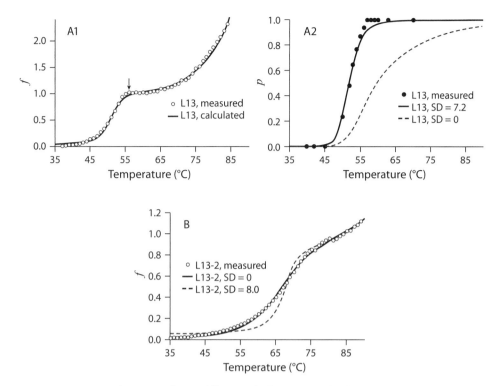

FIGURE 2.11. Melting curves for two different 13-bp-long DNA molecules (L13 and L13-2). Circles represent experimental measurements; lines are fits using the model described in the next section (for (A1) and (A2), identical parameter values were used in the model). Adapted from Ivanov, Zeng, and Zocchi (2004).

(A1): The measurements are obtained by UV absorption spectroscopy, which reports on a combination of base pairing and base stacking. The data are normalized so that the signal is $f = 0$ well below the melting transition and $f = 1$ at the strand dissociation point. In the region $0 \leq f < 1$ the normalized spectroscopic measure f represents essentially the fraction of unpaired bases in the sample. For $f > 1$, it represents essentially the degree of unstacking in the sample.

(A2): The same melting transition as in (A1), observed by the quenching method described in the text. This method measures the fraction, p, of dissociated molecules. The dotted line shows what happens in the model if the strand dissociation entropy (SD) is set to zero.

(B): UV absorption profile for a different 13mer, L13-2, which, unlike L13, is not partially self-complementary. The discrepancy between the values of the strand dissociation entropy SD necessary to obtain good fits for A and B highlights quantitative deficiencies in the model used.

extreme behavior in figure 2.11(A1) is that the oligomer sequence (L13: length 13 bp) is almost completely self-complementary. This has two effects: it pushes the strand separation transition to lower temperature (through the existence of cruciform states), and it enhances stacking in the ss through base-pairing interactions in the single strand.

A more usual melting profile is the one in figure 2.11(B), obtained for a sequence of the same length but no self-complementarity. Strand separation is complete at $T \approx 75\,C$, and there is still a prominent unstacking process for $T > 75\,C$. Clearly, statistical mechanics models should incorporate both pairing and stacking degrees of freedom in order to be useful to experimentalists.

Another obvious conclusion from the experimental melting profiles is that, for the DNA nanorod, a two-states description of melting is wholly inadequate. Figure 2.12 shows melting curves of two different oligomers, of lengths 60 and 42 bp. The quantity f (open circles) represents UV absorption measurements. Normalized to 1 at the strand dissociation point, in the region $f < 1, f$ reports essentially the fraction of unpaired bases (as the midpoint of the unstacking transition is at relatively higher temperature, $> 100\,C$). The quantity p (filled circles) is the fraction of dissociated molecules, measured with the quenching method just described. If the transition was two states (either the entire molecule is dissociated, or no bases are unpaired), we would have $f = p$ for the entire range $f \leq 1$. The measurements prove that even for short molecules, for $T < T_c$ a fraction of the molecules populate "intermediate," partially melted states. The specific sequences of figure 2.12 were designed with AT-rich regions in the middle, of length 36 (L60B36: total length 60, length of bubble-forming region 36) and 18 (L42B18), respectively. The ends are "clamped" by GC-rich regions. In this case, the partially melted part is in the middle of the molecule, contrary to the premise of the zipper model. From the two melting curves $f(T)$ and $p(T)$ one obtains two further quantities. The fraction of open bp at temperature T is

$$f(T) = [1 - p(T)]\langle \ell \rangle + p(T), \qquad (2.41)$$

where $\langle \ell \rangle(T)$ is the average fractional length of the melted region (bubble) in the partially melted molecules (i.e., averaged over the subset of the partially open molecules). Thus the fractional bubble length is

$$\langle \ell \rangle = \frac{f - p}{1 - p}, \qquad (2.42)$$

and this quantity is plotted as squares in figure 2.12. The $\langle \ell \rangle$ vs. T curves for different lengths of the AT-rich region are reminiscent of the isotherms in the P–V plane for a liquid–gas transition, where plateaus signify phase coexistence (of bubbles and paired states, in our case).

The second quantity we can calculate from f, p is simply

$$\sigma = f - p, \qquad (2.43)$$

which represents the fraction of bases in a bubble state. Figure 2.13 shows this quantity, plotted vs. $T - T_c$, for a series of molecules of decreasing lengths, from 48 to 13 bp. The sequences are designed to open from one end (i.e., there

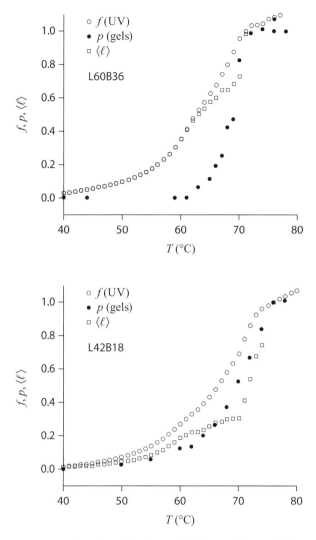

FIGURE 2.12. Experimental melting profiles for two different DNA molecules: L60B36 has total length 60 bp and a bubble-forming region of length 36 in the middle, and correspondingly for L42B18. Open circles are the normalized UV absorption measurements $f(T)$; they represent the fraction of open bp. Filled circles are measurements with the quenching method $p(T)$; they represent the fraction of dissociated molecules. Squares represent the relative length of the bubble, $\langle \ell \rangle$, and are calculated from (2.42). Adapted from Zeng, Montrichok, and Zocchi (2003).

is an AT-rich region at one end). The plot shows that even for the 13mer, the melting transition is not two states. A more quantitative statement is obtained by plotting (figure 2.14) σ_{av}, which is the area under the σ curve of figure 2.13

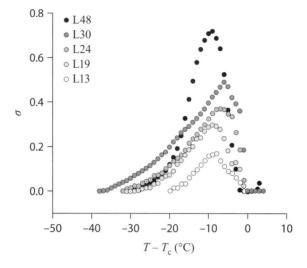

FIGURE 2.13. Experimental measurements of the fraction of bases in a bubble state, σ, plotted vs. $T - T_c$, where T_c is the strand dissociation temperature. The quantity σ is obtained from (2.43) using the experimental values of f and p. This series of molecules, of lengths from 48 bp down to 13 bp, have an AT-rich region at one end. Adapted from Zeng, Montrichok, and Zocchi (2004).

divided by the width of the peak. The quantity σ_{av} measures the frequency of intermediate (partially melted) states averaged over the transition region. The plot of σ_{av} vs. L (the length of the molecule) extrapolates basically to the origin. It shows that the melting transition is strictly two states only for molecules of length $L = 1$!

In summary, experiments on the melting of oligomers show that

1. a two-states description is inadequate;
2. both pairing and stacking degrees of freedom must be considered;
3. cooperativity in this transition is more subtle than is captured by the simple zipper model.

2.5 Base Pairing and Base Stacking as Separate Degrees of Freedom

One can capture the essentials of the melting curve of oligomers in the whole accessible temperature range (figure 2.11) through simple, analytically solvable, statistical mechanics models which give insight into the physics of

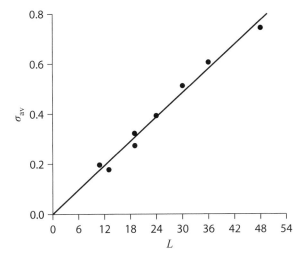

FIGURE 2.14.   The quantity σ_{av} is the area under the peaks of figure 2.13 divided by the width of the peak, here plotted against the length of the molecule, L. This quantity is a measure of the occurrence of intermediate states, averaged over the transition region. The linear fit (solid line) indicates that the transition is truly two states only for molecules of length ~ 1 bp! Adapted from Zeng, Montrichok, and Zocchi (2004).

these molecules. We start again with the zipper model for the base-pairing degrees of freedom, described by the partition sum

$$Z_{\text{pairing}} = \sum_{n=0}^{N-1} (n+1)e^{n(s-\varepsilon/T)} + e^{N(s-\varepsilon/T)}S_{d}. \qquad (2.44)$$

As in (2.33), this partition sum considers only partially melted states where the base pairs are open contiguously from the ends. The factor $(n + 1)$ counts states open from both ends of the molecule. The last $(n = N)$ term includes the dissociation entropy S_d, which depends on DNA concentration.

The stacking degrees of freedom are very well described by an Ising-type model of noninteracting dipoles in a magnetic field. With r stacking bonds that might be broken, the partition sum is

$$Z_{\text{stacking}} = [1 + e^{(\sigma-\Delta/T)}]^{r}. \qquad (2.45)$$

Each unstacking causes an entropy gain $\sigma > 0$ and energy loss $\Delta > 0$, and (2.45) describes a completely uncooperative process where each stacking degree of freedom is independent. Physically, it says that bases can flip out of stack without causing big distortions of the overall structure, which would couple to the next stack. The broad, uncooperative transition predicted by (2.45) fits the experimental measurements very well.

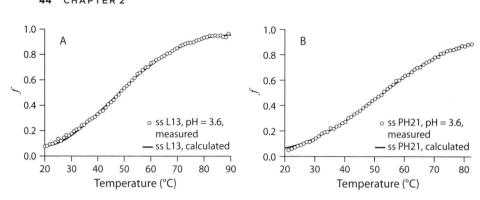

FIGURE 2.15. Unstacking melting curves obtained by UV absorption measurements on single-stranded molecules, of (left) length 13 and (right) length 21 bases. The broad, uncooperative transition is well described by an Ising model of noninteracting spins (solid line, eq. (2.46)). Adapted from Ivanov, Zeng, and Zocchi (2004).

Figure 2.15 shows unstacking profiles obtained from UV absorption of single strands, of length 13 and 21 bases. Using the stratagem of lowering the pH (the data were taken at pH = 3.6), the midpoint of the unstacking transition, which is above 100 C at neutral pH, has been brought down to ~ 50 C, making the whole melting profile accessible. The solid lines are fits of the experimental data using the fraction of unstacked bases f_u given by the model (2.45), that is,

$$f_u = \frac{1}{e^{(\Delta/T - \sigma)} + 1}. \tag{2.46}$$

Now we combine the base-pairing description (2.44) with the base-stacking description (2.45), enforcing the following crucial constraint: bases can be unstacked only if they are unpaired. This is a geometric constraint reflecting the structure of the double helix.

Consider then an oligomer of length N, with a and b open base pairs at the two ends, $a + b = n$, $n < N$; the number of stacking bonds that might be broken is $2n$. Therefore,

$$Z = \sum_{n=0}^{N-1} (n+1) e^{n(s - \varepsilon/T)} [1 + e^{(\sigma - \Delta/T)}]^{2n} + e^{N(s - \varepsilon/T)} [1 + e^{(\sigma - \Delta/T)}]^{2N-2} S_d$$

$$= Z_b + Z_d \tag{2.47}$$

(bubble states + dissociated state). Since it is essentially a geometric series, the partition sum (2.47) can be summed in closed form using the formulas

$$\sum_{n=0}^{N} x^n = \frac{1 - x^{N+1}}{1 - x}, \qquad \sum_{n=0}^{N} n x^n = \frac{x}{(1-x)^2} [1 - (N+1)x^N + Nx^{N+1}]. \tag{2.48}$$

With the notation

$$x := e^{s-\varepsilon/T}[1 + e^{\sigma-\Delta/T}]^2, \tag{2.49}$$

the sum in (2.47) adds up to

$$Z_{\rm b} = \frac{1}{(1-x)^2}[1 - (N+1)x^N + Nx^{N+1}]. \tag{2.50}$$

The probability that n base pairs per molecule are open (within the subset of partially open molecules) is

$$p(n) = \frac{1}{Z_{\rm b}}(n+1)e^{n(s-\varepsilon/T)}[1 + e^{\sigma-\Delta/T}]^{2n} \tag{2.51}$$

and the average bubble length is (supposing there is one contiguous bubble)

$$\langle \ell \rangle = \langle n \rangle = \sum_{n=0}^{N-1} np(n) = \frac{\partial}{\partial s} \ln Z_{\rm b}. \tag{2.52}$$

To compare with the experimental melting curves, we assume that the UV absorption signal $f(T)$, normalized ($f = 1$ at the dissociation temperature $T_{\rm c}$, $f = 0$ for $T \to 0$), is a linear combination of unpairing and unstacking contributions:

$$f = a\frac{\langle n \rangle_{\rm tot}}{N} + \beta\frac{\langle r \rangle_{\rm tot}}{2(N-1)}, \tag{2.53}$$

where n is the fraction of unpaired bases and r the fraction unstacked; a and β are optical absorption parameters. On the other hand, the fraction p of dissociated molecules is directly measured by the quenching method, and in terms of the model we have

$$p = \frac{Z_{\rm d}}{Z}. \tag{2.54}$$

The total number of unpaired bases and unstacked bases per molecule in the sample are

$$\langle n \rangle_{\rm tot} = \frac{\partial \ln Z}{\partial s}, \qquad \langle r \rangle_{\rm tot} = \frac{\partial \ln Z}{\partial \sigma}. \tag{2.55}$$

It may be convenient to exhibit the contribution from the dissociated molecules separately, by writing

$$\langle n \rangle_{\rm tot} = \langle n \rangle \frac{Z_{\rm b}}{Z} + N\frac{Z_{\rm d}}{Z}, \tag{2.56}$$

with $\langle n \rangle$ given by (2.52), and similarly,

$$\langle r \rangle_{\rm tot} = \frac{\partial \ln Z_{\rm b}}{\partial \sigma}\frac{Z_{\rm b}}{Z} + \frac{2N-2}{e^{(\Delta/T-\sigma)}}\frac{Z_{\rm d}}{Z}. \tag{2.57}$$

The solid lines in figure 2.11 represent the model {(2.47), (2.53)–(2.55)}, with some technical modifications to deal with the specific sequences. It is clear that the model satisfactorily captures the distinct contributions of base pairing and base stacking. On the other hand, the cooperativity enforced by the zipper model for pairing is too extreme, as indicated in figure 2.11(B).

2.6 Hamiltonian Formulation of the Zipper Model

To understand the extreme cooperativity encoded in the partition function (2.33), it is useful to formulate the zipper model through the following hierarchical Hamiltonian. We consider N binary degrees of freedom $\{\varphi_1, \varphi_2, \ldots, \varphi_N\}$ which take the values $\varphi_i \in \{0, 1\}$. A value $\varphi_i = 1$ means the ith base is paired, whereas $\varphi_i = 0$ means it is unpaired. The Hamiltonian is

$$H = -\varepsilon[\varphi_1 + \varphi_1\varphi_2 + \varphi_1\varphi_2\varphi_3 + \cdots + \varphi_1 \cdots \varphi_N]. \tag{2.58}$$

The ground state is $E = -N\varepsilon$ (all bases paired: $\varphi_i = 1$ for all i). If $\varphi_1 = 0$, then each term in the sum (2.58) is zero, independent of the values of the other variables, so $E = 0$. Similarly, if the first p bases are paired ($\varphi_1 = \varphi_2 = \cdots = \varphi_p = 1$) and $\varphi_{p+1} = 0$, then $E = -p\varepsilon$, independent of the state of the other variables φ_i with $i > p + 1$. There are 2^{N-p-1} such states. We see that the Hamiltonian (2.58) represents the zipper model (2.33) with $s = \ln 2$ ($g = 2$). The structure of (2.58) is hierarchical: φ_1 is the most important variable, φ_N the least important. The interaction range is the system size; this is the origin of the phase transition for $N \to \infty$. With finite-range interactions, for example the Ising model in zero field,

$$H = -\varepsilon[\varphi_1\varphi_2 + \varphi_2\varphi_3 + \cdots + \varphi_{N-1}\varphi_N], \tag{2.59}$$

there is no phase transition in the thermodynamic limit.

Coming back to cooperativity, in terms of Ising-like Hamiltonians, the two extreme cases are the all-or-none transition,

$$H = -\varepsilon[\varphi_1\varphi_2 \cdots \varphi_N], \tag{2.60}$$

and the independent variables,

$$H = -\varepsilon[\varphi_1 + \varphi_2 + \cdots + \varphi_N]. \tag{2.61}$$

Obviously one can formulate any degree of cooperativity in between in this manner.

2.7 2 × 2 Model: Cooperativity from Local Rules

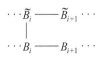

FIGURE 2.16. Dimer for the 2 × 2 model. Bases B_i and $\widetilde{B}_i$ represent complementary bases on opposite strands; B_i and B_{i+1} are adjacent bases on the same strand. Vertical lines represent base-pairing bonds, horizontal lines base-stacking bonds. Adapted from Ivanov, Piontkovski, and Zocchi (2005).

Now we study a model with separate base-pairing and base-stacking degrees of freedom, but where cooperativity of base pairing arises from a more physical, local rule (interactions are not infinitely long range as in (2.58)). This discussion is also an exercise in the transfer matrix formalism of statistical mechanics.

We construct the states of the DNA molecule using a series of overlapping "dimers" (figure 2.16). We denote by B_i the ith base on one strand, counted say from the 5′ end; $\widetilde{B}_i$ is the complementary base on the other strand, counted from the 3′ end (the two strands in the double helix are antiparallel).

The vertical lines in the figure represent base-pairing bonds, the horizontal lines base-stacking bonds. The ith dimer contains the i and $(i + 1)$ pairings and stackings as shown. Breaking these bonds has free energy costs G_i^{p}, G_i^{st}, and $\widetilde{G}_i^{\mathrm{st}}$ for the pairing between B_i & $\widetilde{B}_i$, the stacking between B_i & B_{i+1}, and the stacking between $\widetilde{B}_i$ & $\widetilde{B}_{i+1}$, respectively. We also introduce corresponding statistical weights $U_i^{\mathrm{p}} = e^{-G_i^{\mathrm{p}}/T}$, $U_i^{\mathrm{st}} = e^{-G_i^{\mathrm{st}}/T}$, and $\widetilde{U}_i^{\mathrm{st}} = e^{-\widetilde{G}_i^{\mathrm{st}}/T}$. A dimer has 4 bonds (2 pairings, 2 stackings), each of which can be closed or open (broken), so there are $2^4 = 16$ possible states of the dimer. These are represented in the diagrams of figure 2.17, where vertical lines represent pairing bonds, horizontal lines stacking bonds, and crosses signify broken bonds. The partition function for the model is obtained by summing the statistical weights for all the diagrams for a series of dimers that represents the sequence of the DNA molecule. Now we introduce certain local rules that reflect geometric constraints in the molecular structure and result in cooperative behavior intermediate between independent dipoles in a magnetic field (no cooperativity) and the zipper model (full cooperativity). The rules are that, in the dimer, stacking bonds can be broken only if one pairing bond is broken. Also, if exactly one pairing is broken, then at least one stacking must be broken. In terms of the diagrams of figure 2.17, the rules mean that diagrams 12 to 16 are not allowed: these are crossed out in the figure. These rules implicitly assign a free energy penalty for opening a bubble, because they require mandatory unstacking at a melting fork. The geometrical origin of the constraints is that base pairing can prevent unstacking (while the reverse is not true), while unpairing one bp, but not the next, requires that the bases in the first pair are spatially separated, while in the second pair they are not. This is impossible without at least one unstacking.

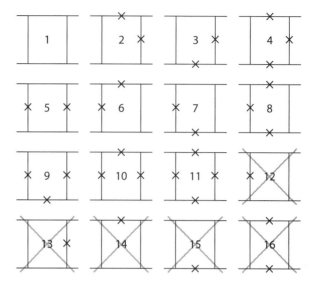

FIGURE 2.17. Dimer states for the 2×2 model. Vertical lines represent base-pairing bonds, horizontal lines base-stacking bonds. A cross indicates a broken bond. The 2×2 model is obtained by disallowing diagrams 12 to 16, which are crossed out. Adapted from Ivanov, Piontkovski, and Zocchi (2005).

Now we write the partition function for this model using the transfer matrix formalism, which is especially convenient for dealing with nonhomogeneous sequences. The idea of the transfer matrix is to write the partition sum for the system of size $N + 1$ in terms of the partition sum for the system of size N, basically setting up a recursion relation. In our case, let us call Z_N^c the partition sum for the system (of N bp) with the last bp closed (i.e., paired), and use Z_N^o for the last bp open. The total partition sum (neglecting, for the moment, end effects) is

$$Z_N = Z_N^c + Z_N^o. \tag{2.62}$$

Then we can write

$$\begin{cases} Z_{N+1}^c = Z_N^c U_1 + Z_N^o [U_6 + U_7 + U_8], \\ Z_{N+1}^o = Z_N^c [U_2 + U_3 + U_4] + Z_N^o [U_5 + U_9 + U_{10} + U_{11}], \end{cases} \tag{2.63}$$

where U_1 is the statistical weight corresponding to diagram 1 in figure 2.17, etc. For example, diagrams 6, 7, 8 all contribute to the "transition" o $\rightarrow$ c, diagrams 2, 3, 4 to the transition c $\rightarrow$ o, and so on. The statistical weights in question are

$$
\begin{cases}
U_1 = 1, \quad U_6 = \sqrt{U_i^{\mathrm{p}}}\, \tilde{U}_i^{\mathrm{st}}, \quad U_7 = \sqrt{U_i^{\mathrm{p}}}\, U_i^{\mathrm{st}}, \quad U_8 = \sqrt{U_i^{\mathrm{p}}}\, U_i^{\mathrm{st}} \tilde{U}_i^{\mathrm{st}}, \\[4pt]
U_2 = \sqrt{U_{i+1}^{\mathrm{p}}}\, \tilde{U}_i^{\mathrm{st}}, \quad U_3 = \sqrt{U_{i+1}^{\mathrm{p}}}\, U_i^{\mathrm{st}}, \quad U_4 = \sqrt{U_{i+1}^{\mathrm{p}}}\, U_i^{\mathrm{st}} \tilde{U}_i^{\mathrm{st}}, \\[4pt]
U_5 = \sqrt{U_i^{\mathrm{p}} U_{i+1}^{\mathrm{p}}}, \quad U_9 = \sqrt{U_i^{\mathrm{p}} U_{i+1}^{\mathrm{p}}}\, U_i^{\mathrm{st}}, \quad U_{10} = \sqrt{U_i^{\mathrm{p}} U_{i+1}^{\mathrm{p}}}\, \tilde{U}_i^{\mathrm{st}}, \\[4pt]
U_{11} = \sqrt{U_i^{\mathrm{p}} U_{i+1}^{\mathrm{p}}}\, U_i^{\mathrm{st}} \tilde{U}_i^{\mathrm{st}}.
\end{cases}
\tag{2.64}
$$

The partition function represented by the recursion relation (2.63) is a sum over dimers; each pairing interaction is therefore counted twice (except for the first and last bp): the ith bp occurs in the $(i-1)$th and the ith dimer. Therefore each free energy of pairing G_i^{p} should be halved, and correspondingly the statistical weights for pairing U_i^{p} appear under square root in (2.64). The relation (2.63) can be written

$$
\begin{pmatrix} Z_{N+1}^{\mathrm{c}} \\ Z_{N+1}^{\mathrm{o}} \end{pmatrix} = A_N \begin{pmatrix} Z_N^{\mathrm{c}} \\ Z_N^{\mathrm{o}} \end{pmatrix},
\tag{2.65}
$$

where the transfer matrices A_i ($i = 1, 2, \ldots, N-1$, for a molecule of N bp) are

$$
A_i = \begin{pmatrix}
1 & \sqrt{U_i^{\mathrm{p}}}\,(U_i^{\mathrm{st}} + \tilde{U}_i^{\mathrm{st}} + U_i^{\mathrm{st}} \tilde{U}_i^{\mathrm{st}}) \\
\sqrt{U_{i+1}^{\mathrm{p}}}\left(U_i^{\mathrm{st}} + \tilde{U}_i^{\mathrm{st}} + U_i^{\mathrm{st}} \tilde{U}_i^{\mathrm{st}}\right) & \sqrt{U_i^{\mathrm{p}} U_{i+1}^{\mathrm{p}}}\left(1 + U_i^{\mathrm{st}} + \tilde{U}_i^{\mathrm{st}} + U_i^{\mathrm{st}} \tilde{U}_i^{\mathrm{st}}\right)
\end{pmatrix}.
\tag{2.66}
$$

Now (2.65) can be written

$$
\begin{pmatrix} Z_N^{\mathrm{c}} \\ Z_N^{\mathrm{o}} \end{pmatrix} = A_{N-1} A_{N-2} \cdots A_2 \begin{pmatrix} Z_2^{\mathrm{c}} \\ Z_2^{\mathrm{o}} \end{pmatrix},
\tag{2.67}
$$

where Z_2 is the partition function for the system consisting of 2 bp (1 dimer): Z_2^{c} the partition sum for the dimer with the last (second) bp closed, Z_2^{o} for the last bp open. We now have to introduce boundary conditions (b.c.), and we will use free b.c. (the bp's at the beginning and the end of the molecule can be either closed or open: they are not clamped either way). The column vector

$$
Y = \begin{pmatrix} Y_1 \\ Y_2 \end{pmatrix} \equiv \begin{pmatrix} Z_1^{\mathrm{c}} \\ Z_1^{\mathrm{o}} \end{pmatrix}
\tag{2.68}
$$

which implements these b.c., that is, such that

$$
\begin{pmatrix} Z_2^{\mathrm{c}} \\ Z_2^{\mathrm{o}} \end{pmatrix} = A_1 \begin{pmatrix} Y_1 \\ Y_2 \end{pmatrix},
\tag{2.69}
$$

is

$$
Y_1 = 1, \qquad Y_2 = \sqrt{U_1^{\mathrm{p}}},
\tag{2.70}
$$

as we can see from (2.68). Indeed, using (2.70) and (2.69) we find

$$
\begin{cases}
Z_2^c = 1 + U_1^P \left(U_1^{st} + \tilde{U}_1^{st} + U_1^{st}\tilde{U}_1^{st} \right), \\
Z_2^o = \sqrt{U_2^P} \left(U_1^{st} + \tilde{U}_1^{st} + U_1^{st}\tilde{U}_1^{st} \right) + U_1^P \sqrt{U_2^P} \left(1 + U_1^{st} + \tilde{U}_1^{st} + U_1^{st}\tilde{U}_1^{st} \right),
\end{cases}
\tag{2.71}
$$

whereas calculating Z_2^c, Z_2^o directly from the diagrams of figure 2.17, we find that with free b.c., the contributing diagrams are

$$
\begin{cases}
Z_2^c \text{ (b.c.)} = U_1 + U_6 + U_7 + U_8, \\
Z_2^o \text{ (b.c.)} = U_2 + U_3 + U_4 + U_5 + U_9 + U_{10} + U_{11}.
\end{cases}
\tag{2.72}
$$

For the statistical weights in (2.72), U_1^P comes without square root because the first pairing is counted only once in the partition sum. Then (2.72) is the same as (2.71).

Similarly, the free b.c. at the end of the molecule is implemented by the row vector

$$
X = \left(X_1 \ X_2 \right) = \left(1 \ \sqrt{U_N^P} \right)
\tag{2.73}
$$

such that the total partition sum is

$$
Z_N = \left(1 \ \sqrt{U_N^P} \right) \begin{pmatrix} Z_N^c \\ Z_N^o \end{pmatrix}.
\tag{2.74}
$$

Finally, the partition sum with free b.c. takes the form

$$
Z_N = \left(1 \ \sqrt{U_N^P} \right) A_{N-1} A_{N-2} \cdots A_1 \begin{pmatrix} 1 \\ \sqrt{U_1^P} \end{pmatrix},
\tag{2.75}
$$

with the transfer matrices A_i given in terms of the unpairing and unstacking free energies by (2.66). For a homogeneous sequence, all the matrices A_i are the same, and symmetric. The transfer matrix can be diagonalized, and the partition sum written in closed form; the thermodynamic limit $N \to \infty$ can also be calculated: this is one way to solve the 1-D Ising model, for example. But in general, with (2.75) and (2.66) we can describe any finite sequence. One caution, however, is that the complete set of 16 unstacking free energies for the 16 possible distinct stacking interactions in the Watson–Crick paired dimer has not been measured. To see that there are actually, in principle, 16 different possible stacking interactions between the 4 bases is not immediately obvious. It is best to write down the 16 different dimers, and recognize which out of the 32 stacking bonds are equivalent. For example, $\begin{smallmatrix}3'-T\ G\\5'-A\ C\end{smallmatrix}$ and $\begin{smallmatrix}3'-C\ A\\5'-G\ T\end{smallmatrix}$ are equivalent dimers; in $\begin{smallmatrix}3'-T\ A\\5'-A\ T\end{smallmatrix}$ the two stacking bonds are equivalent, and so on.

Coming back to the partition sum, we still have to include a dissociation entropy S_d with the term corresponding to all bp being open. First we calculate the statistical weight of the collection of states with all pairings open: only diagrams $5, 9, 10, 11$ contribute, therefore the A matrices have the form (see (2.63), (2.64))

$$A_i^{\text{diss}} = \begin{pmatrix} 0 & 0 \\ 0 & \sqrt{U_i^{\text{p}} U_{i+1}^{\text{p}}} \left(1 + U_i^{\text{st}} + \tilde{U}_i^{\text{st}} + U_i^{\text{st}} \tilde{U}_i^{\text{st}} \right) \end{pmatrix} \tag{2.76}$$

and the required statistical weight is

$$D = \left(0 \ \sqrt{U_N^{\text{p}}} \right) A_{N-1}^{\text{diss}} \cdots A_1^{\text{diss}} \begin{pmatrix} 0 \\ \sqrt{U_1^{\text{p}}} \end{pmatrix}$$
$$= \sqrt{U_N^{\text{p}} U_1^{\text{p}}} \prod_{i=1}^{N-1} \sqrt{U_i^{\text{p}} U_{i+1}^{\text{p}}} \left(1 + U_1^{\text{st}} + \tilde{U}_1^{\text{st}} + U_1^{\text{st}} \tilde{U}_1^{\text{st}} \right). \tag{2.77}$$

This statistical weight is present in Z_N given by (2.75), so to isolate the dissociation term we write the total partition sum thus:

$$Z_{\text{tot}} = (Z_N - D) + e^{S_d} D. \tag{2.78}$$

The term in parentheses contains all the partially melted states, and the last term all the dissociated states. Measurable quantities are calculated as follows. We refer to the subset of partially melted molecules, and define the column vectors

$$B_i = A_{i-1} A_{i-2} \cdots A_1 \begin{pmatrix} 1 \\ \sqrt{U_1^{\text{p}}} \end{pmatrix} \quad (2 \le i \le N); \quad B_1 \equiv \begin{pmatrix} 1 \\ \sqrt{U_1^{\text{p}}} \end{pmatrix} \tag{2.79}$$

and the row vectors

$$C_i = \left(1 \ \sqrt{U_N^{\text{p}}} \right) A_{N-1} A_{N-2} \cdots A_i \quad (1 \le i \le N - 1); \quad C_N \equiv \left(1 \ \sqrt{U_N^{\text{p}}} \right). \tag{2.80}$$

Then the partition sum is

$$Z_N = C_i B_i = c_1^i b_1^i + c_2^i b_2^i. \tag{2.81}$$

Equation (2.81) is valid for any i ($1 \le i \le N$), there is no summation over i, and we have introduced the components

$$C_i = \begin{pmatrix} c_1^i & c_2^i \end{pmatrix}, \quad B_i = \begin{pmatrix} b_1^i \\ b_2^i \end{pmatrix}. \tag{2.82}$$

In (2.81), the first term on the right-hand side is the statistical weight that the ith base is closed, the second term the statistical weight for the ith base being open. So the probability that the ith base is open (within the subset of non-dissociated molecules) is

$$P_i(\text{open}) = \frac{1}{Z_N - D} c_2^i b_2^i \tag{2.83}$$

and the fraction of open bp is

$$f = \sum_{i=1}^{N} P_i(\text{open}). \tag{2.84}$$

Similarly, let us find the probability that the ith stacking (on the principal strand) is broken. We write A_i with contributions from diagrams 3, 4, 7, 8, 9, 11 only:

$$A_i^{\text{st}} = \begin{pmatrix} 0 & \sqrt{U_i^{\text{p}}} \left(U_i^{\text{st}} + U_i^{\text{st}} \tilde{U}_i^{\text{st}} \right) \\ \sqrt{U_{i+1}^{\text{p}}} \left(U_i^{\text{st}} + U_i^{\text{st}} \tilde{U}_i^{\text{st}} \right) & \sqrt{U_i^{\text{p}} U_{i+1}^{\text{p}}} \left(U_i^{\text{st}} + U_i^{\text{st}} \tilde{U}_i^{\text{st}} \right) \end{pmatrix}. \tag{2.85}$$

The required probability is

$$P_i^{\text{st}}(\text{open}) = \frac{1}{Z_N - D} C_{i+1} A_i^{\text{st}} B_i \quad (1 \le i \le N - 1). \tag{2.86}$$

Figure 2.18 shows experimental melting profiles and fits using the 2×2 model. It is presently not known whether one consistent set of parameters for the 2×2 model can be found such that the melting profiles for any sequence are reproduced to the accuracy shown in figure 2.18.

The 2×2 model is conceptually appealing because it describes, in a transparent manner, the interplay of the most relevant degrees of freedom—pairing and stacking—in shaping the melting curves of DNA oligomers. For practical applications such as predicting melting temperatures of oligomers, models with a reduced number of parameters are desirable, since as we mentioned before, only some of the 16 different stacking interactions have actually been measured. Commonly used is the nearest neighbor (NN) model, described now.

2.8 Nearest Neighbor Model

The NN model is obtained from the 2×2 model by lumping together stacking and pairing interactions into effective pairing interactions for the different dimers. That is, we remove the stacking degrees of freedom altogether, and we correspondingly renormalize the pairing free energies. If G_i^{NN} is the effective free energy cost of opening the ith bp, and $U_i^{\text{NN}} = e^{-G_i^{\text{NN}}/T}$, (2.66) now takes the form

$$A_i = \begin{pmatrix} 1 & \sqrt{U_i^{\text{NN}}} \\ \sqrt{U_{i+1}^{\text{NN}}} & \sqrt{U_i^{\text{NN}} U_{i+1}^{\text{NN}}} \end{pmatrix}. \tag{2.87}$$

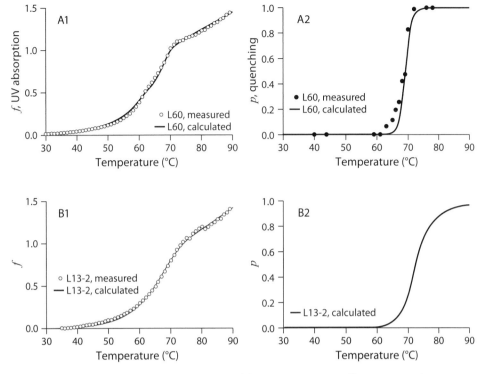

FIGURE 2.18. Melting profiles and the 2×2 model. Top: measurements of base-pairing and base-stacking $f(T)$ obtained by UV absorption (A1), and the dissociation curve $p(T)$ obtained from the quenching method, for a DNA molecule of length 60 bp. The lines are fits with the 2×2 model (same parameter values for (A1) and (A2)).
Bottom: same for a DNA molecule of length 13 bp. Adapted from Ivanov, Piontkovski, and Zocchi (2005).

In terms of the 2×2 model parameters, all 16 diagrams in figure 2.17 are supposed to contribute to the transitions in (2.63), for example, for the o → c transition, diagrams 6, 7, 8, 12 contribute, and therefore,

$$U_i^{\mathrm{NN}} = U_i^{\mathrm{p}}\left(U_i^{\mathrm{st}} + \widetilde{U}_i^{\mathrm{st}} + U_i^{\mathrm{st}}\widetilde{U}_i^{\mathrm{st}} + 1\right) = U_i^{\mathrm{p}}\left(1 + U_i^{\mathrm{st}}\right)\left(1 + \widetilde{U}_i^{\mathrm{st}}\right). \qquad (2.88)$$

This is a transparent result: $\left(1 + U_i^{\mathrm{st}}\right)$ is the partition sum describing stacking for the primary strand of the dimer, $\left(1 + \widetilde{U}_i^{\mathrm{st}}\right)$ for the complementary strand. In terms of free energies,

$$G_i^{\mathrm{NN}} = G_i^{\mathrm{p}} - T\ln(1 + e^{-G_i^{\mathrm{st}}/T}) - T\ln(1 + e^{-\widetilde{G}_i^{\mathrm{st}}/T}) \qquad (2.89)$$

and using $G = H - TS$, $\partial G/\partial T = -S$ we can calculate, from (2.89), the temperature dependence of the NN parameters H_i^{NN}, S_i^{NN} in terms of the (temperature-independent) 2×2 model parameters H_i, S_i, where $G_i^{\mathrm{st}} = H_i^{\mathrm{st}} - TS_i^{\mathrm{st}}$ etc. Now from (2.88) it is easy to see by direct calculation that the partition sum for the NN model is simply

$$Z^{NN} = \left(1 \ \sqrt{U_N^{NN}}\right) A_{N-1}^{NN} A_{N-2}^{NN} \cdots A_1^{NN} \begin{pmatrix} 1 \\ \frac{1}{\sqrt{U_1^{NN}}} \end{pmatrix}$$

$$= \left(1 + U_1^{NN}\right)\left(1 + U_2^{NN}\right) \cdots \left(1 + U_N^{NN}\right). \tag{2.90}$$

Unless we add by hand a penalty for opening bubbles, this is again a completely uncooperative model of independent dipoles in a magnetic field. For example, the probability that the ith bp is open is

$$P_i^{NN}(\text{open}) = \frac{U_i^{NN}}{1 + U_i^{NN}}. \tag{2.91}$$

So the NN model is obtained from the 2×2 model by relaxing the geometrical constraints corresponding to the forbidden diagrams in figure 2.17 (and lumping stacking and pairing degrees of freedom together). On the other hand, the zipper model is obtained from the 2×2 model by *adding* more constraints, to ensure that base pairs open only contiguously. The one-sided zipper model (allowing the molecule only to open contiguously from one side) with stacking degrees of freedom included, that is, essentially model (2.47) apart from the multiplicity factor $(n + 1)$, is obtained from the 2×2 model by forbidding diagrams 2, 3, 4, in addition to 13, 14, 15, 16; diagram 12, on the other hand, is allowed. (Referring to figure 2.17, we are allowing the zipper to open "from the left".) Then the transfer matrix takes the form

$$A_i = \begin{pmatrix} 1 & \sqrt{U_i^P}\left(1 + U_i^{st}\right)\left(1 + \tilde{U}_i^{st}\right) \\ 0 & \sqrt{U_i^P U_{i+1}^P}\left(1 + U_i^{st}\right)\left(1 + \tilde{U}_i^{st}\right) \end{pmatrix}. \tag{2.92}$$

We can see this describes a zipper model in that $\begin{pmatrix} 1 \\ 0 \end{pmatrix}$ is an eigenvector of (2.92), therefore if the ith bp is closed (i.e., open with statistical weight 0), the $(i + 1)$th bp is also closed. If now we remove the stacking degrees of freedom, as in (2.87), we obtain the simple zipper model

$$A_i^{zip} = \begin{pmatrix} 1 & \sqrt{U_i} \\ 0 & \sqrt{U_i U_{i+1}} \end{pmatrix}. \tag{2.93}$$

For a homogeneous sequence ($U_i = U$ for all i) it is easy to see by direct calculation that the partition sum is

$$Z_N^{zip} = \left(1 \ \sqrt{U}\right) A_{N-1}^{zip} A_{N-2}^{zip} \cdots A_1^{zip} \begin{pmatrix} 1 \\ \frac{1}{\sqrt{U}} \end{pmatrix} = \sum_{n=0}^{N} U^n, \tag{2.94}$$

which is the same as (2.33).

In summary, the 2×2 model addresses the role of the most relevant degrees of freedom for the DNA nanorod, and represents a conceptual framework for a whole class of Ising-type models of DNA melting. Melting curves in this context play the role of specific heat curves in more general condensed matter systems, in that they tell us what kind of excitations of the ground state structure are important. Reduced degrees of freedom models of these excitations, such as the Debye model of phonons in solids, are indeed a traditional means of gaining insight into the physics of condensed matter systems.

2.9 Connection to Nonlinear Dynamics

DNA conformational transitions also represent an intriguing problem in nonlinear physics, and this connection is best exemplified by the Peyrard–Bishop–Dauxois (PBD) model of DNA melting. It is a 1-D model where the nth base pair is associated with one continuous variable $y_n(t)$, representing the deviation from the equilibrium distance between the two complementary bases on the opposite strand; $p_n(t) = m\dot{y}_n$ is the associated momentum. The Hamiltonian of the model is

$$H = \sum_n \frac{p_n^2}{2m} + D\left(e^{-ay_n} - 1\right)^2 + \frac{K}{2}\left[1 + \rho e^{-\beta(y_n+y_{n-1})}\right](y_n - y_{n-1})^2. \quad (2.95)$$

The first term is the kinetic energy of the bp. The second term describes the Watson–Crick base-pairing interaction: breaking a base pairing ($y \gg 1/a$) has an energy cost D. The last term describes base stacking. The sum is over all base pairs (however, for a molecule of N bp there are only $N-1$ base-stacking terms).

The stacking term consists of a harmonic potential $\propto (y_n - y_{n-1})^2$ multiplied by a nonlinear term (the square bracket) which has the following effect. When the two adjacent base pairs are "closed" ($y_n = y_{n-1} = 0$), the strength of the harmonic potential is $K[1 + \rho]$. When one of the adjacent base pairs, or both, are significantly displaced ($y_n > 1/\beta$ or $y_{n-1} > 1/\beta$), the strength of the base-stacking interaction is reduced to K. It is this nonlinear term that is responsible for the remarkable ability of the model to quantitatively describe DNA melting. Without the term in the square bracket, the Hamiltonian (2.95) gives rise to a melting transition that is too soft compared to real DNA; the nonlinear term introduces essentially a change in stiffness between paired and unpaired tracts of the molecule, which translates into an additional entropy increase upon melting. Indeed, one interesting feature of the model is the embedded connection between nonlinearity and entropy generation. Parameter values for (2.95) that apply to DNA are $\rho = 2$, $1/a \approx 0.2$ Å, $D \approx 2$ kT, $K \approx 1$ kT/Å^2, $1/\beta \approx 3$ Å.

For a real DNA molecule the sequence would typically be inhomogeneous, and that is accommodated in the model by making the parameters D

and a dependent on the specific bp (AT or GC), that is, $D \rightarrow D_n$ and $a \rightarrow a_n$ in (2.95), with two possible values depending on the identity of the bp. The equilibrium statistical mechanics generated by (2.95) is most easily explored by Langevin dynamics simulations; the surprise is that this relatively simple model, with one set of optimized parameters, reproduces the melting curves of different DNA sequences very well. In contrast, the Ising-type models of the preceding sections typically require some adjustment to the parameters when applied to different sequences. On the other hand, the PBD model does not describe the experimental melting curves beyond the strand separation temperature, because it does not allow for residual stacking in the single strands.

In summary, oligonucleotide melting is a problem that can also be discussed in an interesting fashion from the perspective of nonlinear physics.

2.10 Linear and Nonlinear Elasticity of DNA

The next interesting conformational property of the DNA nanorod is its ability to bend reversibly. One implication is that short DNA molecules can be used as leaf springs for nanotechnology applications. The fundamental quantity of interest is the elastic energy of bending. What is measured in experiments is invariably a *free* energy of bending, and one interesting question is whether entropic contributions to this quantity are important. This question naturally leads to the study of the temperature dependence of the elastic properties of the molecule. The boldest and, it turns out, most effective approach to DNA elasticity is to adopt an unashamedly continuum mechanics viewpoint and describe DNA bending as the bending of a rod. This viewpoint allows us to explore, quantitatively, bending deformations of DNA nanorods in the whole energy range from linear to nonlinear response.

2.11 Bending Modulus and Persistence Length

For a uniform thin rod and small bending, the elastic energy of bending is

$$E = \int_0^L ds \, \frac{1}{2} B \kappa^2(s), \tag{2.96}$$

where $\kappa(s)$ is the curvature, s is the arclength (and $\kappa(s) = 1/R(s)$, where R is the radius of curvature), and L is the contour length of the (thin) rod. The parameter B (dimensions of energy $\times$ length $=$ force $\times$ length^2), which describes the bending stiffness of the rod, is called the bending modulus. For a cylindrical rod,

$$B = Y \frac{1}{4} \pi r^4, \tag{2.97}$$

where Y is Young's modulus and r is the radius of the cylinder. On the other hand, in polymer physics one describes the stiffness of a long polymer through its persistence length ℓ_p, which is roughly speaking the length of polymer over which thermal fluctuations cause significant bending. Evidently ℓ_p and B are related quantities; consider a polymer of

FIGURE 2.19. Tangent vectors to a polymer chain.

length ℓ_p: we ask that the energy cost of bending it into a half circle be of order T. The radius of curvature is then $R = \ell_p/\pi$, that is, $\kappa = \pi/\ell_p$; using (2.96) the bending energy is $E = \ell_p(\frac{1}{2})B(\pi/\ell_p)^2$ and the condition $E \sim T$ gives

$$\ell_p \sim \frac{B}{T}, \tag{2.98}$$

forgetting about the numerical factor in this essentially dimensional argument.

It is instructive to be more precise about the relation between the bending modulus B and the persistence length ℓ_p. First we need a precise definition of ℓ_p, which is obtained by considering correlation functions. Let $\vec{n}(s)$ be the unit tangent vector to the contour of the rod at position s (figure 2.19).

Due to thermal fluctuations, the correlation function $\langle \vec{n}(0) \cdot \vec{n}(s) \rangle$ decays exponentially with arclength s:

$$\left\langle \vec{n}(0) \cdot \vec{n}(s) \right\rangle = e^{-s/\ell_p}, \tag{2.99}$$

where the decay length is the persistence length ℓ_p. The brackets $\langle \; \rangle$ denote an ensemble average. Now we prove (2.99) and find ℓ_p in terms of B. In terms of the unit tangent vector $\vec{n}(s)$, the scalar curvature is

$$\kappa(s) = \left| \frac{d\vec{n}}{ds} \right|. \tag{2.100}$$

We consider a plane curve for simplicity (then our result for ℓ_p will be valid in 2-D); representing 2-D vectors as points in the complex plane we can write

$$\vec{n} = e^{i\theta(s)}, \tag{2.101}$$

where θ is the angle that the tangent to the rod makes with the real axis; then,

$$\frac{d\vec{n}}{ds} = i\frac{d\theta}{ds}\,\vec{n} \quad \Rightarrow \quad \kappa = \left| \frac{d\vec{n}}{ds} \right| = \left| \frac{d\theta}{ds} \right| \tag{2.102}$$

or

$$\kappa(s) = \frac{d\theta}{ds} \tag{2.103}$$

if we keep the sign in the curvature. We choose coordinates such that $\theta(0) = 0$; then $\vec{n}(0) \cdot \vec{n}(s) = \cos[\theta(s) - \theta(0)] = \cos[\theta(s)]$, that is,

$$\langle \vec{n}(0) \cdot \vec{n}(s) \rangle = \langle \cos[\theta(s)] \rangle. \tag{2.104}$$

To calculate the expectation value in (2.104), we first use an approximate, but simple, scheme which, however, gives the exact result. Namely, we consider only bent states that are in the shape of an arc of a circle, that is, shapes for which the curvature κ is a constant: $\kappa(s) = a = \text{const.}$ For a segment of length s and curvature a, the energy is $E = (\frac{1}{2})sBa^2$; we write the partition sum as

$$Z = \int_0^\infty da \exp\left(-\tfrac{sB}{2}a^2/T\right). \tag{2.105}$$

Since from (2.103) and the boundary condition $\theta(0) = 0$ we have $\theta(s) = as$, the correlation function (2.104) is

$$\langle \vec{n}(0) \cdot \vec{n}(s) \rangle = \langle \cos(as) \rangle = \frac{1}{Z} \int_0^\infty da \, \cos(as) \exp\left(-\tfrac{sB}{2T}a^2\right). \tag{2.106}$$

The integrals in (2.105), (2.106) are obtained from the formula

$$\int_0^\infty e^{-ax^2} \cos(bx)\, dx = \frac{1}{2}\sqrt{\frac{\pi}{a}} \exp\left(-\tfrac{b^2}{4a}\right) \tag{2.107}$$

and we find

$$\langle \vec{n}(0) \cdot \vec{n}(s) \rangle = \exp\left(-\tfrac{T}{2B}s\right), \tag{2.108}$$

showing that the correlation function does indeed decay exponentially, with a decay length

$$\ell_{\mathrm{p}} = \frac{2B}{T} \quad \text{(2-D)}, \tag{2.109}$$

which is the exact result in 2-D. In 3-D, the exact result is different by a factor 2:

$$\ell_{\mathrm{p}} = \frac{B}{T} \quad \text{(3-D)}. \tag{2.110}$$

Now, for illustration purposes, we can redo this calculation without restricting ourselves to shapes of constant curvature. For a given shape $\theta(s)$, the energy is

$$E = \frac{1}{2}B \int_0^\Delta ds \left(\frac{d\theta}{ds}\right)^2 \tag{2.111}$$

(see (2.103)); Δ is the contour length of the rod. The physical rod is $0 \leq s \leq \Delta$, but we can formally extend $\theta(s)$ as an odd function in the interval $-\Delta \leq s \leq \Delta$ and develop $\theta(s)$ in a Fourier series using only cosines:

$$\theta(s) = \sum_{n=0}^\infty b_n \cos\left(\tfrac{\pi}{\Delta}ns\right). \tag{2.112}$$

The boundary condition $\theta(0) = 0$ gives

$$b_0 + \sum_{n=1}^{\infty} b_n = 0, \tag{2.113}$$

while

$$\theta(\Delta) = b_0 + \sum_{n=1}^{\infty} (-1)^n b_n. \tag{2.114}$$

Substituting b_0 from (2.113) into (2.114) we find

$$\theta(\Delta) = -2 \sum_{n=0}^{\infty} b_{2n+1} \tag{2.115}$$

$$\Rightarrow \quad \theta^2(\Delta) = 4 \sum_{n,m=0}^{\infty} b_{2n+1} b_{2m+1}. \tag{2.116}$$

We now write the energy in terms of the b degrees of freedom:

$$\frac{d\theta}{ds} = -\frac{\pi}{\Delta} \sum_{n=1}^{\infty} n b_n \sin\left(\frac{\pi}{\Delta} n s\right), \tag{2.117}$$

$$E = \frac{1}{2} B \left(\frac{\pi}{\Delta}\right)^2 \sum_{n,m=1}^{\infty} n m b_n b_m \int_0^{\Delta} ds \, \sin\left(\frac{\pi}{\Delta} n s\right) \sin\left(\frac{\pi}{\Delta} m s\right). \tag{2.118}$$

The integral is $(\Delta/2)\delta_{nm}$, and we obtain

$$E = \frac{\pi^2 B}{4\Delta} \sum_{n=1}^{\infty} n^2 b_n^2, \tag{2.119}$$

showing that the b's are independent degrees of freedom, and therefore $\langle b_i b_j \rangle = \langle b_i^2 \rangle \delta_{ij}$ so that from (2.116) we find

$$\langle \theta^2(\Delta) \rangle = 4 \sum_{n=0}^{\infty} \left\langle b_{2n+1}^2 \right\rangle. \tag{2.120}$$

By equipartition we find from (2.119),

$$\frac{\pi^2 B}{4\Delta} n^2 \left\langle b_n^2 \right\rangle = \frac{1}{2} T \quad \Rightarrow \quad \left\langle b_n^2 \right\rangle = \frac{2 T \Delta}{\pi^2 B} \frac{1}{n^2}, \tag{2.121}$$

and finally

$$\langle \theta^2(\Delta) \rangle = 4 \frac{2 T \Delta}{\pi^2 B} \sum_{n=0}^{\infty} \frac{1}{(2n+1)^2}. \tag{2.122}$$

The sum is $\pi^2/8$, so that

$$\langle \theta^2(\Delta) \rangle = \frac{T}{B} \Delta. \tag{2.123}$$

Now we can write

$$\langle \cos[\theta(\Delta)] \rangle \approx 1 - \tfrac{1}{2}\langle \theta^2(\Delta) \rangle \approx \exp\left[-\tfrac{1}{2}\langle \theta^2(\Delta) \rangle\right] \tag{2.124}$$

for small θ, and therefore we obtain once again

$$\langle \cos[\theta(\Delta)] \rangle = \langle \vec{n}(0) \cdot \vec{n}(\Delta) \rangle = \exp\left(-\tfrac{T}{2B}\Delta\right). \tag{2.125}$$

2.12 Measurements of DNA Elasticity: Long Molecules

The persistence length in polymer physics is traditionally measured from the size of long polymer coils in dilute solution, determined by neutron, X-ray, or light scattering. For DNA, mechanical stretching experiments on single molecules introduced by Bustamante and coworkers are now routinely employed to determine the bending modulus B under various conditions. In these experiments, one end of a long ($\sim 10\,\mu$m) DNA molecule is attached to a surface, the other end to a micron-size bead. The bead is manipulated using either a flexible cantilever or by optical tweezers, or, in the case of a paramagnetic bead, through a magnetic field gradient. "Manipulated" means that the external force on the bead and the position of the bead can be recorded. One obtains force–extension curves for the molecular tether (the DNA molecule). The worm-like chain (WLC) model, which is based on the elastic energy (2.96), is then used to extract the value of B from the experimental measurements.

In more detail, consider a long DNA molecule, of contour length $L \gg \ell_{\rm p}$; one end is attached at the origin, the other end, located by the coordinates (x, y, z), is pulled by a constant applied force f, which we take to be in the z-direction. In equilibrium, the molecule attains an end-to-end distance or extension $\langle z \rangle$. Call $\vec{n}(s)$ the unit vector tangent to the contour of the DNA, where s is the position along the DNA ($0 \le s \le L$). In terms of the field $\vec{n}(s)$, the potential energy part of the Hamiltonian for the worm-like-chain model is

$$H[\vec{n}(s)] = \tfrac{1}{2}B \int_0^L ds \left(\frac{\partial \vec{n}}{\partial s}\right)^2 - f \int_0^L ds\, n_3(s), \tag{2.126}$$

where $\vec{n} = (n_1, n_2, n_3)$ and $|\partial_s \vec{n}| = \kappa(s)$ is the curvature ($|\vec{n}| = 1$). The negative sign in the second term favors fluctuations of the extension

$$z = \int_0^L ds\, n_3(s) \tag{2.127}$$

in the direction of the applied force. The force–extension relation (f vs. $\langle z \rangle$) is obtained from

$$\langle z \rangle = L \langle n_3 \rangle, \tag{2.128}$$

where the ensemble average is calculated using (2.126).

Calculating the partition sum for the model (2.126) is not simple; the solution was given in 1995 by Marko and Siggia. However, the two limits $\langle z \rangle \ll L$ and $\langle z \rangle \approx L$ are relatively simple. Taking the second first, we assume $n_3 \approx 1$, $n_1, n_2 \ll 1$ for all s and divide the unit tangent vector $\vec{n}$ into a component along the applied force and an orthogonal component $\vec{n}_o$:

$$\vec{n} = \vec{n}_o + n_3 \hat{z}. \tag{2.129}$$

Since $|\vec{n}_o| \ll 1$ we have

$$n_3 = \sqrt{1 - |\vec{n}_o|^2} \approx 1 - \tfrac{1}{2}|\vec{n}_o|^2, \tag{2.130}$$

and also

$$\left| \frac{\partial n_3}{\partial s} \right| \ll \left| \frac{\partial n_1}{\partial s} \right|, \left| \frac{\partial n_2}{\partial s} \right| \quad \Rightarrow \quad \left(\frac{\partial \vec{n}}{\partial s} \right)^2 \approx \left(\frac{\partial \vec{n}_o}{\partial s} \right)^2, \tag{2.131}$$

so that the Hamiltonian (2.126) may be written in terms of $\vec{n}_o(s)$:

$$H = \frac{1}{2} \int_0^L ds \left\{ B \left(\frac{\partial \vec{n}_o}{\partial s} \right)^2 + f(\vec{n}_o)^2 \right\} - fL \tag{2.132}$$

and (2.128) becomes

$$\langle z \rangle = L \left(1 - \tfrac{1}{2} \left\langle |\vec{n}_o|^2 \right\rangle \right). \tag{2.133}$$

The two components of $\vec{n}_o = (n_1, n_2)$ being independent degrees of freedom in this approximation, we may consider the scalar version of (2.132), with a scalar field $n_o(s)$, and at the end use

$$\left\langle |\vec{n}_o|^2 \right\rangle = 2 \left\langle n_o^2 \right\rangle. \tag{2.134}$$

Expanding $n_o(s)$ in normal modes,

$$n_o(s) = \sum_{n=0}^{\infty} a_n \sin \left(\frac{n\pi}{L} s \right), \qquad \frac{\partial n_o}{\partial s} = \sum_{n=0}^{\infty} \frac{n\pi}{L} a_n \cos \left(\frac{n\pi}{L} s \right), \tag{2.135}$$

and substituting into (2.132) we find

$$H = \frac{1}{2} \sum_{n,m} \frac{n\pi}{L} \frac{m\pi}{L} a_n a_m B \int_0^L ds \cos \left(\frac{n\pi}{L} s \right) \cos \left(\frac{m\pi}{L} s \right)$$

$$+ \frac{1}{2} \sum_{n,m} a_n a_m f \int_0^L ds \sin \left(\frac{n\pi}{L} s \right) \sin \left(\frac{m\pi}{L} s \right) - fL. \tag{2.136}$$

The integrals are $(L/2)\delta_{n,m}$, so finally,

$$H = \frac{L}{4} \sum_{n=0}^{\infty} \left[\left(\frac{n\pi}{L} \right)^2 B + f \right] a_n^2 - fL. \tag{2.137}$$

The Hamiltonian is quadratic in the amplitudes a_n and therefore by equipartition we have

$$\frac{L}{4}\left[\left(\frac{n\pi}{L}\right)^2 B + f\right]\langle a_n^2 \rangle = \tfrac{1}{2}T$$

$$\Rightarrow \quad \langle a_n^2 \rangle = \frac{2T}{L}\frac{1}{(\frac{n\pi}{L})^2 B + f}, \tag{2.138}$$

while

$$\langle n_o^2 \rangle = \frac{1}{L}\int_0^L ds \sum_{n,m}\langle a_n a_m \rangle \sin\left(\frac{n\pi}{L}s\right)\sin\left(\frac{m\pi}{L}s\right)$$

$$= \frac{1}{2}\sum_{n=0}^{\infty}\langle a_n^2 \rangle . \tag{2.139}$$

Approximating the sum with an integral,

$$\sum_{n=0}^{\infty}\frac{1}{(\frac{n\pi}{L})^2 B + f} \approx \int_0^{\infty} du\,\frac{1}{\frac{\pi^2 B}{L^2}u^2 + f} = \frac{L}{\pi\sqrt{fB}}\int_0^{\infty} dv\,\frac{1}{v^2 + 1}. \tag{2.140}$$

The last integral is $\pi/2$, and we obtain

$$\langle n_o^2 \rangle = \frac{1}{2}\frac{2T}{L}\frac{L}{2}\frac{1}{\sqrt{fB}} = \frac{1}{2}\frac{T}{\sqrt{fB}}, \tag{2.141}$$

and from (2.133),

$$\frac{\langle z \rangle}{L} = 1 - \frac{1}{2}\frac{T}{(fB)^{1/2}} = 1 - \frac{1}{(4f\ell_p/T)^{1/2}}, \tag{2.142}$$

where in the last equality we have used the relation between bending modulus B and persistence length ℓ_p. Inverting (2.142) we find the force–extension relation for large extension,

$$f = \frac{T}{4\ell_p}\frac{1}{(1 - \langle z \rangle/L)^2}. \tag{2.143}$$

In the opposite limit, of small extension, the applied force is a perturbation that does not alter the statistics of the chain, which for zero force is Gaussian (chapter 1). The entropic force of the Gaussian chain at extension z is

$$f = \frac{3T}{\langle R^2 \rangle}z, \tag{2.144}$$

where $\langle R^2 \rangle$ is the average end-to-end distance squared at zero force. If correlations along the chain decrease exponentially (true in particular for the WLC),

$$\langle \vec{n}(s)\cdot\vec{n}(s') \rangle = e^{-|s-s'|/\ell_p}, \tag{2.145}$$

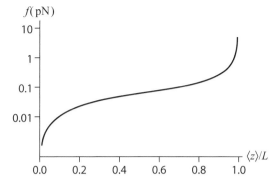

FIGURE 2.20. Plot of eq. (2.149) using $\ell_p = 50\,\text{nm}$, i.e., $T/\ell_p = (4/50)\,\text{pN}$.

with ℓ_p the persistence length, then the second moment of the end-to-end distance

$$\vec{R} = \int_0^L ds\, \vec{n}(s) \tag{2.146}$$

is

$$\left\langle |\vec{R}|^2 \right\rangle = \int_0^L ds \int_0^L ds'\, \langle \vec{n}(s) \cdot \vec{n}(s') \rangle$$
$$= \int_0^L ds \int_0^L ds'\, e^{-|s-s'|/\ell_p} = 2L\ell_p + 2\ell_p^2(e^{-L/\ell_p} - 1), \tag{2.147}$$

the last equality after a little algebra. For $L \gg \ell_p$ we have therefore $\langle |\vec{R}|^2 \rangle = 2L\ell_p$ and

$$f = \frac{3T}{2L\ell_p} \langle z \rangle \tag{2.148}$$

for small extension. The Marko–Siggia expression,

$$f = \frac{T}{\ell_p} \left[\frac{1}{4(1 - \langle z \rangle /L)^2} + \frac{\langle z \rangle}{L} - \frac{1}{4} \right] \tag{2.149}$$

is a useful interpolation formula that is exact for the two limits above (as we can easily find out), and a very good approximation in between. Figure 2.20 shows the worm-like-chain force–extension curve according to (2.149).

A most interesting phenomenon occurs in the experiments as the stretching force is further increased in the regime $\langle z \rangle \approx L$. At about $f = 70\,\text{pN}$ the force–extension curve enters a plateau region, where the DNA molecule can be "overstretched" from $\langle z \rangle = L$ to about $\langle z \rangle = 1.7L$, at essentially constant force (figure 2.21). This is the first example we encounter of a reversible softening or yield transition in the mechanics of hydrogen-bonded molecular

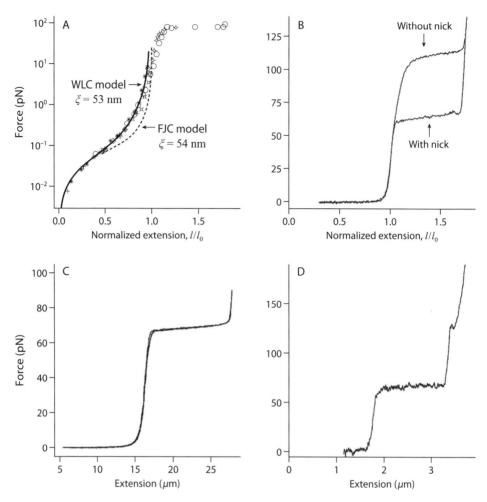

FIGURE 2.21. Experimental force–extension curves for different DNA molecules, measured by different groups with different force transducers. The plateau at ∼ 70 pN signals a yield transition called DNA overstretching. WLC: worm-like chain; FJC: freely jointed chain. Reproduced from Strick et al. (2000).

structures. Such softening transitions turn out to be fairly ubiquitous in the mechanics of biomolecules from DNA to enzymes, and indeed represent one guiding thread for the materials science approach proposed in this book. Similar to order–disorder transitions in different fields of condensed matter physics, the microscopic, "structural" basis for these softening transitions may be quite disparate for different systems, such as kinking of a DNA nanorod and "induced-fit" conformational motion of an enzyme. There is, however, a degree of universality in the thermodynamic behavior, for instance the stress–strain relations, so that it is useful to think about these

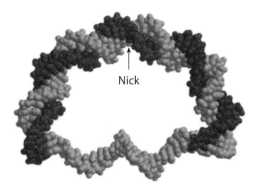

FIGURE 2.22. D-DNA: this molecule consists of two
linear strands hybridized as shown. These constructs
are used to measure the elastic energy of highly bent
DNA. The figure is a composite cartoon, adapted from
Qu and Zocchi (2011).

systems in terms of simple models that reproduce the measured thermodynamic behavior. This is the approach we take throughout the book.

The single molecule experiments on DNA are in the tradition of polymer
physics experiments, in that they probe the mechanics of very long, flexible
molecules. Nonetheless, they also led to the discovery of the overstretching
transition, which is within the class of molecular deformation phenomena
that form the subject of this book. We now come back to the nanoscale.

2.13 Measurements of DNA Elasticity: Short Molecules

We consider bending deformations of a DNA nanorod of contour length
L ($L \ll \ell_{\mathrm{p}}$). Ideally, one wants to measure the elastic energy for different
degrees of bending. This is achieved through a surprisingly simple experiment. Consider the DNA molecule ("D-DNA") shown schematically in
figure 2.22: it is formed by two linear (i.e., not circular) strands of different
lengths, with the region of complementarity arranged as shown (the two
ends of the lighter strand are complementary to the two halves of the darker
strand). Thus there is by construction a nick in the middle of the ds region
of the molecule (the sugar–phosphate backbone of the lighter strand is
interrupted there). This arrangement is essential for the measurement that
follows. The molecule of figure 2.22 consists of the DNA nanorod (the ds part
of the molecule) and a linear spring (the ss part) resulting from the stretching
elasticity of ss DNA. The nanorod is forced to bend due to the constraint of
the ss part pulling on its ends. The whole molecule is, by construction, under
stress. We will call N_{d} the number of base pairs in the ds part, and N_{s} the
number of bases in the ss part. The following observation provides a method

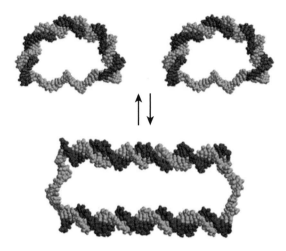

FIGURE 2.23. Dimerization mechanism for D-DNA with a nick. Adapted from Sanchez et al. (2013).

to measure the elastic energy of the molecule. Two D-DNA molecules ("monomers") can come together to form a dimer (figure 2.23). One dimer contains exactly the same Watson–Crick base pairing as two monomers. However, mechanical stress in the dimer is relaxed. Dimer formation *is driven* by the release of stress of the monomer. The elastic energy of the monomer can then be measured from the fraction of monomers (M) and dimers (D) at equilibrium in the "reaction" $2M \rightleftharpoons D$. Namely, writing the chemical potentials for the monomer and dimer as (see (2.15))

$$\begin{cases} \mu_M = \mu_M^0 + E_{el} + T \ln X_M, \\ \mu_D = \mu_D^0 + T \ln X_D, \end{cases} \tag{2.150}$$

where E_{el} is the elastic energy of the monomer, the equilibrium condition $2\mu_M - \mu_D = 0$ gives

$$E_{el} = \frac{1}{2} \ln \frac{X_D}{X_M^2} = \frac{1}{2} \ln \left(C_w \frac{[D]}{[M]^2} \right) \tag{2.151}$$

(see (2.17)), since $2\mu_M^0 - \mu_D^0 \approx 0$; this last condition is due to the fact that two monomers and one dimer are essentially identical molecules from the viewpoint of internal bonds, surface exposed to water, etc.

The equilibrium concentrations appearing in (2.151) can be measured, for example, by gel electrophoresis (figure 2.24). Measuring the elastic energy (2.151) for a series of molecules with varying N_s, fixed N_d, one measures in effect the bending energy of the nanorod for various degrees of bending. Figure 2.25 shows such energy curves for two different nanorods, of length

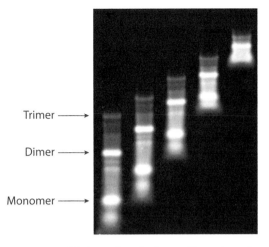

FIGURE 2.24. Gel electrophoresis of an equilibrated sample of D-DNA. The overall DNA concentration was chosen such that both monomers and dimers are clearly visible and can be quantified from the intensity of the bands. All lanes contain the same sample, loaded at successive times. Adapted from Qu et al. (2011).

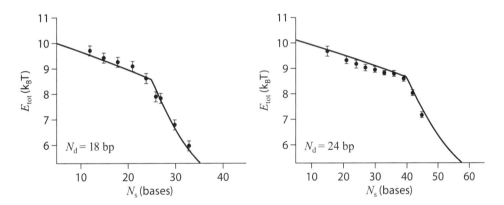

FIGURE 2.25. Measurement of the total elastic energy E_{tot} for a series of D-DNA molecules with $N_d = 18$ (left) and $N_d = 24$ (right), and varying N_s. The lines are calculated using the model (2.206). The "kink" in these energy profiles signals a softening transition which corresponds to the appearance of a kink in the DNA nanorod.

$N_d = 18$ and $24\,\mathrm{bp}$ ($L = 6$ and $8\,\mathrm{nm}$). Of course, the measurements represent the total elastic energy E_{tot} of the molecule of figure 2.22, which is the sum of the bending energy of the nanorod and the stretching energy of the ss portion. Nonetheless, it is obvious that, as N_s is reduced (i.e., for increasing bending of the nanorod), there is a softening or yield transition

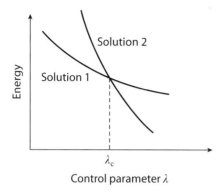

FIGURE 2.26. Two different branches of the energy curve give rise to a bifurcation.

in the mechanical response. For the shorter nanorod ($N_d = 18$), the "yield point" is at $N_s = N_c \approx 25$; for the longer rod, $N_c \approx 40$. It turns out that for $N_s > N_c$ the nanorod is within the linear elasticity regime described by the worm-like-chain elastic energy (2.96), while for $N_s < N_c$ the nanorod has developed a reversible, constant torque, kink. We now investigate this interesting nonlinear mechanical response quantitatively.

2.14 The Euler Instability

The energy curves in figure 2.25 suggest an instability or *bifurcation* phenomenon. In mechanical terms, this is a situation where there are *two* solutions for the equilibrium state, one stable and one unstable; as a control parameter is varied, the stable solution becomes unstable and vice versa.

In terms of energy curves, there are two branches, and they cross at the critical value of the control parameter (figure 2.26). The stable solution is the one of lower energy: as λ is increased, the system jumps from solution 1 to solution 2. This is a very important and general scenario in physics, encompassing phase transitions of matter and bifurcations of dynamical systems. It is intellectually pleasing to find the same phenomenon in the mechanics of a molecule, because the scales are so different. This is the driving concept of this section.

The most famous bifurcation—the first to be analyzed quantitatively, by Euler around 1750—is the Euler instability. As a warm-up exercise, we first discuss this classic problem; then we deploy the same methods to analyze the yield transition of the DNA nanorod.

As everyone knows, a straight rod under a compressive force F buckles "suddenly" at a critical force F_c. This is the Euler instability. Under a compressive force F, there are two mechanical equilibrium solutions (figure 2.27): one is the straight rod, the other is the buckled rod. If the rod is straight, the elastic energy is

$$E_s = \tfrac{1}{2}K(x - L)^2, \tag{2.152}$$

where $K = Y\pi r^2/L$ is the "spring constant" of the straight rod; x is the end-to-end distance, L is the uncompressed length of the rod (for compressive

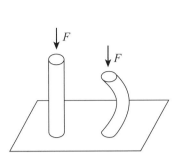

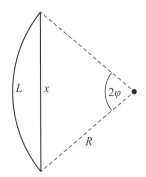

FIGURE 2.27. The Euler instability: buckling of a rod under a compressive force.

FIGURE 2.28. Geometry for the rod of length L bent into an arc of a circle of radius R.

forces, $x \leq L$), Y is Young's modulus, and πr^2 is the cross-sectional area of the rod, supposed circular.

The equilibrium shape of the buckled rod can of course be calculated exactly, but it is nontrivial. Instead, we use an approximate trial shape and find the elastic energy for that shape.

As a first approximation, let us take an arc of a circle of contour length L (figure 2.28). The end-to-end distance (the length of the chord) is x, and the radius of curvature is R. From the figure,

$$\frac{x}{2R} = \sin \varphi, \ 2\varphi R = L \quad \Rightarrow \quad \frac{x}{2R} = \sin\left(\frac{L}{2R}\right). \tag{2.153}$$

For small bending ($L/2R \ll 1$), using (2.153),

$$\frac{x}{2R} \approx \frac{L}{2R} - \frac{1}{6}\left(\frac{L}{2R}\right)^3 \quad \Rightarrow \quad \frac{x-L}{L} \approx -\frac{1}{6}\left(\frac{L}{2R}\right)^2. \tag{2.154}$$

The elastic energy of the bent rod is therefore

$$E_{\mathrm{b}} = \frac{1}{2}B\frac{L}{R^2} = -12\frac{B}{L^2}(x-L) \tag{2.155}$$

in this approximation, using (2.96) and (2.154). As we see, E_{b} is *linear* in $(L-x)$. Thus the two branches of the energy function are

$$E(x) = \begin{cases} \frac{1}{2}Y\frac{\pi r^2}{L}(L-x)^2 & \text{(straight)}, \\ 12\frac{B}{L^2}(L-x) & \text{(bent)}, \end{cases} \tag{2.156}$$

and as we see from figure 2.29, for small compression the stable (lower-energy) solution is the straight one, up to a critical compression where the bent solution becomes the stable one.

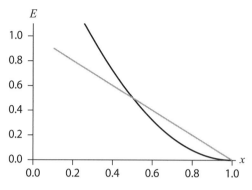

FIGURE 2.29. Plot of eq. (2.156) for $L = 1$, $\frac{1}{2} Y \pi r^2 / L = 2$, $12B/L^2 = 1$.

The critical force for buckling is the derivative of the bent solution at the critical point (this is because normally we choose the applied force F as the control parameter, rather than the end-to-end distance x):

$$F_c = -\frac{\partial E_b}{\partial x}\bigg|_{x=x_c} = 12\frac{B}{L^2}. \tag{2.157}$$

The critical end-to-end distance x_c is found from $E_s(x_c) = E_b(x_c)$:

$$\frac{1}{2}Y\frac{\pi r^2}{L}(L - x_c) = 12\frac{B}{L^2} \quad \Rightarrow \quad x_c = L\left(1 - 6\frac{r^2}{L^2}\right). \tag{2.158}$$

The equilibrium energy function is given by (2.156), where one has to choose the upper expression for $x_c \le x \le L$ and the lower expression for $x \le x_c$. The value of x_c is found by equating the two expressions. Notice that the force $F = -\partial E / \partial x$ is discontinuous at x_c: $F(x_c^-)$ is given by (2.157), while

$$F(x_c^+) = -\frac{\partial E_s}{\partial x}\bigg|_{x=x_c} = Y\frac{\pi r^2}{L}(L - x_c) = 24\frac{B}{L^2}. \tag{2.159}$$

Equation (2.157) is correct except for the numerical coefficient; the exact solution with free boundary conditions (zero torque at the ends of the rod) is

$$F_c = \pi^2\frac{B}{L^2}. \tag{2.160}$$

The reason is of course that the bent shape in reality is not an arc of a circle. In particular, our trial shape does not satisfy the zero-torque boundary conditions at the ends (zero torque means zero curvature), and as a result we considerably overestimate the elastic energy for the bent solution. However, we can improve the calculation by using a trial function for the shape that does satisfy the boundary conditions.

We will use polynomials and work in Cartesian coordinates, assuming small bending. Figure 2.30 shows the geometry: $y = y(x)$ is the shape of the bent rod. The curvature in Cartesian coordinates is

$$\kappa = \frac{d^2y/dx^2}{[1+(dy/dx)^2]^{5/2}} \approx \frac{d^2y}{dx^2}. \qquad (2.161)$$

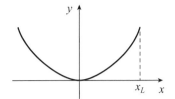

The boundary condition is zero torque at the ends, which means $d^2y/dx^2 = 0$ for $x = \pm x_L$. The shape is symmetric: $y(-x) = y(x)$. The lowest-order polynomial expression that can satisfy the boundary conditions is

FIGURE 2.30. Coordinates for the polynomial solution to the bent rod problem.

$$y = ax^2 - b|x^3|, \quad a, b > 0, \qquad (2.162)$$

the boundary condition imposing

$$2a - 6bx_L = 0 \quad \Rightarrow \quad b = \frac{a}{3x_L}. \qquad (2.163)$$

Our trial shape is therefore parametrized by one constant, a:

$$y = ax^2 - \frac{a}{3x_L}|x^3|, \qquad (2.164)$$

and a is related to x_L by the constraint

$$L = \int ds = \int \sqrt{dx^2 + dy^2} = 2\int_0^{x_L} dx \sqrt{1 + \left(\frac{dy}{dx}\right)^2}. \qquad (2.165)$$

Using (2.164) in (2.165),

$$\begin{aligned}
L &= 2\int_0^{x_L} dx \left[1 + \left(2ax - \frac{a}{x_L}x^2\right)^2\right]^{1/2} \\
&\approx 2\int_0^{x_L} dx \left[1 + \frac{1}{2}\left(2ax - \frac{a}{x_L}x^2\right)^2\right] = 2x_L + \frac{8}{15}a^2x_L^3.
\end{aligned} \qquad (2.166)$$

The end-to-end distance of the rod is $x = 2x_L$. Note that in the small bending approximation $x_L \approx L/2$ and $ax_L \ll 1$, and we are keeping terms up to $O((ax_L)^2)$. The elastic energy of the shape (2.164) is, using (2.161),

$$E = 2 \times \frac{1}{2}B \int_0^{x_L} dx \left(\frac{d^2y}{dx^2}\right)^2 = B\int_0^{x_L} dx \left(2a - \frac{2a}{x_L}x\right)^2 = \frac{4}{3}Ba^2x_L, \qquad (2.167)$$

and, using (2.166),

$$E_b = \frac{5}{2}B\frac{1}{x_L^2}(L - x) \approx 10\frac{B}{L^2}(L - x). \qquad (2.168)$$

The condition $E_b(x_c) = E_s(x_c)$ now gives $x_c = L(1 - 5r^2/L^2)$ (compare with (2.158)) and the critical force is

$$F_c = 10\frac{B}{L^2}. \tag{2.169}$$

Since $\pi^2 \approx 9.86$, the difference between (2.169) and the exact result is less than 2%.

As an aside, it turns out that the exact shape of the rod, in the small bending approximation, is a sine function. Using $y(x) = A[1 - \cos(kx)]$ as the trial shape and performing the same calculation one finds

$$k = \frac{\pi}{2x_L}, \quad L - 2x_L = \frac{\pi^2}{8}\frac{A^2}{x_L}, \quad E_b = \pi^2\frac{B}{L^2}(L - x),$$

$$x_c = L\left(1 - \frac{\pi^2}{2}\frac{r^2}{L^2}\right), \tag{2.170}$$

and the exact result for F_c.

2.15 The DNA Yield Transition

The elastic energy curves in figure 2.25 are essentially linear to the left of the transition point ($N_s < N_c$). This fact suggests that in this regime, the DNA nanorod behaves like a leaf spring of constant force (independent of end-to-end distance x). To see this, consider the construct of figure 2.22 as two coupled springs. The elastic energy is the sum

$$E_{tot} = E_d(x_{eq}) + E_s(x_{eq}), \tag{2.171}$$

where E_d is the (bending) elastic energy of the ds part of the molecule, E_s is the (stretching) elastic energy of the ss part, and x_{eq} is the end-to-end distance corresponding to mechanical equilibrium:

$$\frac{\partial E_d}{\partial x} + \frac{\partial E_s}{\partial x} = 0 \quad \text{for } x = x_{eq}. \tag{2.172}$$

As a first approximation for $E_s(x)$ we take the entropic elasticity of the ideal chain,

$$E_s(x) = \frac{3T}{2N_k\ell_k^2}x^2 = \frac{9T}{4N_s\ell_s^2}x^2, \tag{2.173}$$

where N_k is the number of Kuhn lengths and $\ell_k = 2\ell_s$ is the Kuhn length; ℓ_s is the persistence length. For ss DNA, $\ell_s \approx 0.8$ nm so $\ell_k \approx 1.6$ nm or ~ 5 bases; we assume $N_k = N_s/6$, consistent with the Marko–Siggia formula discussed later. Now suppose a linear dependence for $E_d(x)$:

$$E_d(x) = E_0 - ax, \tag{2.174}$$

that is, a constant force spring. From (2.171)–(2.174),

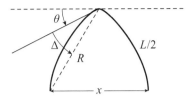

FIGURE 2.31. Geometry for the kinked rod.

$$x_{eq} = a\frac{2\ell_s^2 N_s}{9T} \text{ and } E_{tot} = E_0 - \frac{a^2\ell_s^2}{9T}N_s, \quad (2.175)$$

that is, a linear dependence of E_{tot} on N_s. The experimental measurements of the elastic energy of the molecules (figure 2.25) thus lead to the following model. As the compressive force on the nanorod (in the experiments, the pulling of the ss part) is increased (in the experiments, by reducing N_s) the nanorod bends smoothly according to linear elasticity up to a critical point where the internal bending torque reaches a critical value τ_c. At that point there is a bifurcation to a solution where the nanorod has a kink. With the boundary conditions of the experiments (force applied at the ends; no bending torque at the ends), the internal bending torque in the rod is maximum at the center, so the kink develops at the center. This kink is a constant torque kink (the internal bending torque at the kink is independent of the kink angle, and equal to τ_c). The solid lines in figure 2.25 are calculated with this model, and we can see that they represent the experimental energy curves to within a fraction of $1\,k_BT$. The value of the critical bending torque for a DNA rod *with one nick at the center* is $\tau_c \approx 27\,pN\,nm$, essentially independent of local sequence and of temperature (below the melting transition). The critical bending torque τ_c and the bending modulus $B \approx 200\,pN\,nm^2$ are sufficient to characterize the bending energy of the DNA nanorod in the linear and nonlinear regimes. Considering the complexity of the molecule, this is a remarkable result. The value of τ_c for the case without nick has been measured only indirectly; perhaps the safe statement is that it is $\geq 31\,pN/nm$.

Now we solve the model. Figure 2.31 shows the geometry in the kinked state: the internal angle of the kink is $(\pi - 2\theta)$, while θ is the angle that the tangent to the rod at the kink makes with the "horizontal"; this is different from the angle that the chord R makes with the horizontal, which is $\theta + \Delta$, because the half-rod, of contour length $L/2$, is bent.

For $\theta > 0$ (i.e., in the kinked state), Δ is fixed (independent of θ), and the shape of the half-rod is also fixed, by the boundary conditions $\tau = \tau_c$ at the kink end, $\tau = 0$ at the free end. The internal torque is

$$\tau = B\frac{d\theta(s)}{ds}, \quad (2.176)$$

where $\theta(s)$ is the angle of the tangent to the rod as a function of arclength s (to keep notation simple, we also denote by θ the angle shown in figure 2.31; no confusion will ensue). From the geometry,

$$\frac{x}{2} = R\cos(\theta + \Delta) \quad \Rightarrow \quad \theta = -\Delta + \arccos\left(\frac{x}{2R}\right), \tag{2.177}$$

where

$$\Delta = \arccos\left(\frac{x_c}{2R}\right), \tag{2.178}$$

x_c is the critical value of the end-to-end distance x where the kink appears, and (2.177) is valid for $\theta \geq 0$ (i.e., $x \leq x_c$). The energy in the kinked state is

$$E_k = E_0 - \tau_c\Delta + \tau_c\arccos\left(\frac{x}{2R}\right), \tag{2.179}$$

where E_0 is a constant (independent of x); k stands for kinked. The expression (2.179) gives an energy linear in x for small x, since $\arccos(x) \approx \pi/2 - x$ for $x \ll 1$, and thus

$$E_k = \text{const.} - \frac{\tau_c}{2R}x \quad \text{for } \frac{x}{2R} \ll 1. \tag{2.180}$$

To find the constant E_0 in (2.179), we have to match the energy (2.179) with the energy of the smoothly bent solution, at the critical point. We take coordinates as in figure 2.30, consider only half of the symmetric shape ($x \geq 0$), and allow a cusp at $x = 0$. The internal bending torque is

$$\tau = B\frac{d\theta}{ds} \approx B\frac{d^2y}{dx^2} \tag{2.181}$$

in the approximation of small bending. With our boundary conditions, the maximal internal bending torque occurs at $x = 0$,

$$\tau_0 = B\frac{d^2y}{dx^2}\Big|_{x=0}. \tag{2.182}$$

For the shape of the nanorod, we take a polynomial approximation (see (2.162))

$$y = ax^2 - \frac{a}{3x_L}x^3, \tag{2.183}$$

where we have already satisfied the boundary condition $d^2y/dx^2 = 0$ at $x = x_L$ (zero bending torque at the ends). The constant a in (2.183) is related to the bending torque at $x = 0$: since $(d^2y/dx^2)_{x=0} = 2a$ we have

$$a = \frac{\tau_0}{2B}, \tag{2.184}$$

while x_L is given by the condition (see (2.165))

$$\frac{L}{2} = \int_0^{x_L} dx \sqrt{1 + \left(\frac{dy}{dx}\right)^2} \approx \int_0^{x_L} dx \left[1 + \frac{1}{2}\left(2ax - \frac{a}{x_L}x^2\right)^2\right] \tag{2.185}$$

$$\Rightarrow \quad L = 2x_L + \frac{8}{15}a^2x_L^3, \tag{2.186}$$

or also

$$x_L = \frac{L}{2}\left(1 - \frac{1}{15}a^2L^2\right) \tag{2.187}$$

to order $(aL)^2$. Since $aL = L\tau_0/(2B)$ and $\tau_0 \leq \tau_c$ we can see that this calculation is based on the small parameter

$$\gamma = \frac{L\tau_c}{2B}. \tag{2.188}$$

Given $\tau_c \approx 30\,\text{pN nm}$, $B \approx 200\,\text{pN nm}^2$, the condition $\gamma < 1$ is satisfied by DNA shorter than about 15 nm or 45 bp.

For the chord R we have

$$R^2 = x_L^2 + [y(x_L)]^2 = x_L^2\left[1 + \frac{4}{9}a^2x_L^2\right] \tag{2.189}$$

and using (2.187),

$$R \approx \frac{L}{2}\left(1 - \frac{1}{90}a^2L^2\right) \tag{2.190}$$

to order $(aL)^2$. The energy of the bent rod is

$$\begin{aligned}
E_b &\approx 2\int_0^{x_L} dx\,\frac{1}{2}B\left(\frac{d^2y}{dx^2}\right)^2 = B\int_0^{x_L} dx\left(2a - \frac{2a}{x_L}x\right)^2 \\
&= \frac{4}{3}Ba^2x_L = \frac{2}{3}Ba^2L\left(1 - \frac{1}{15}a^2L^2\right),
\end{aligned} \tag{2.191}$$

where we have used (2.187) and b stands for bent. In terms of the bending torque τ_0 (see (2.184)),

$$E_b = \frac{1}{6}\frac{L}{B}\tau_0^2 \tag{2.192}$$

to lowest order. Matching the two solutions (2.179), (2.192) at the critical point ($\tau_0 = \tau_c$, $x = x_c$), and using (2.178), we have

$$E_0 = \frac{1}{6}\frac{L}{B}\tau_c^2. \tag{2.193}$$

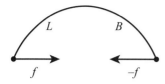

FIGURE 2.32. Bent rod of contour length L under a compressive force f; B is the bending modulus.

From figure 2.31, $\tan(\Delta) = y_c/(x_c/2)$, $y_c = (2/3)a(x_c/2)^2 = (1/12)(\tau_c/B)x_c^2$ and so $\tan(\Delta) = (\tau_c/6B)x_c$, while from (2.187), (2.184), and (2.188),

$$x_c = L\left(1 - \frac{1}{15}\gamma^2\right). \qquad (2.194)$$

In our approximation, Δ is small (of order γ) and therefore

$$\Delta \approx \frac{1}{6}\frac{\tau_c}{B}x_c = \frac{1}{3}\gamma \qquad (2.195)$$

to order γ^2. We see therefore that in expression (2.179), the first two terms on the right-hand side cancel (to order γ^2), and finally, the kinked solution, properly matched to the bent solution at the critical point, has the form

$$E_k = \tau_c \arccos\left(\frac{x}{2R}\right), \qquad (2.196)$$

where, from (2.190) evaluated at the critical point (i.e., $aL = \gamma$),

$$R = \frac{L}{2}\left(1 - \frac{1}{90}\gamma^2\right), \quad \left(\gamma = \frac{L\tau_c}{2B}\right). \qquad (2.197)$$

The complete solution for the elastic energy of the nanorod is therefore (2.192) for $x > x_c$ and (2.196) for $x < x_c$. We can write the former in terms of x, using (2.184) and (2.187):

$$E_b = 10\frac{B}{L^2}(L - x) \qquad (2.198)$$

(see (2.168)). Now, (2.198) is not a satisfactory form for the energy down to $x = L$ ($E_b = 0$). Even if the system were purely mechanical, we know that for $x \to L$ we would run into the Euler instability. More to the point for us, if the elastic energy is of order the temperature or smaller ($E_b \sim T$), a purely mechanical calculation is not appropriate. The Euler instability itself disappears if the critical energy is of order the temperature or smaller. To correct (2.198) for $E_b < T$ we perform an approximate finite T calculation. We consider only flexural modes for which the elastic energy is given by (2.198).

With a compressive force f (see figure 2.32), the total energy vs. end-to-end distance x is

$$E(x) = 10\frac{B}{L^2}(L - x) - f(L - x) = \left(10\frac{B}{L^2} - f\right)(L - x). \qquad (2.199)$$

The scheme is to calculate the ensemble average $\langle x \rangle$ vs. f, invert this relation to find the force f vs. $\langle x \rangle$, and finally obtain the elastic free energy vs. $\langle x \rangle$.

From the partition function

$$Z = -\int_L^0 dx \exp\left[-\frac{1}{T}\left(10\frac{B}{L^2} - f\right)(L - x)\right]$$

$$= \frac{T}{10\frac{B}{L^2} - f}\left\{1 - \exp\left[-\frac{L}{T}\left(10\frac{B}{L^2} - f\right)\right]\right\}, \qquad (2.200)$$

we obtain

$$\langle(L - x)\rangle = \frac{T}{Z}\frac{\partial Z}{\partial f} = \frac{T}{10\frac{B}{L^2} - f} - \frac{L}{\exp\left[\frac{L}{T}\left(10\frac{B}{L^2} - f\right)\right] - 1}. \qquad (2.201)$$

If f is not too large, that is, for the case $f < 10B/L^2$ (e.g., for a DNA 30mer this means $f < 20\,\mathrm{pN}$), the exponential in the last expression is large, and we may keep just the first term,

$$L - x = \frac{T}{10\frac{B}{L^2} - f}, \qquad (2.202)$$

where we denote $\langle x \rangle$ simply by x. Finally,

$$f = 10\frac{B}{L^2} - \frac{T}{L - x} = 10\frac{B}{L^2}\left[1 - \frac{LT}{10B}\frac{1}{1 - x/L}\right], \qquad (2.203)$$

where f is the force; the work done by this force, which is the elastic free energy, is

$$E(x) = -\int_{x_0}^x f\,dx = -10\frac{B}{L^2}(x - x_0) - T\ln\left(\frac{L - x}{L - x_0}\right), \qquad (2.204)$$

where $x_0 < L$ is the (ensemble-averaged) end-to-end distance at zero force, which is less than L because of thermal fluctuations. From (2.203) we see that

$$x_0 = L\left(1 - \frac{LT}{10B}\right). \qquad (2.205)$$

The free energy (2.204) has a quadratic minimum at x_0, as it should, thus curing the linear dependence of (2.199) for low energy, which is unphysical. Finally, the bending elastic energy of the nanorod vs. end-to-end distance x is, according to this model,

$$E(x) = \begin{cases} -10\dfrac{B}{L^2}(x - x_0) - T\ln\left(\dfrac{L - x}{L - x_0}\right) & \text{for } x \geq x_c, \\[2ex] \tau_c \arccos\left(\dfrac{x}{2R}\right) & \text{for } x < x_c, \end{cases} \qquad (2.206)$$

where L is the contour length of the DNA, x_0 is given by (2.205), R by (2.197), and x_c is obtained by equating the upper and lower expressions

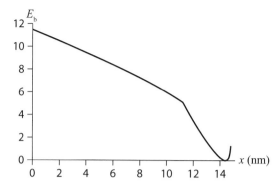

FIGURE 2.33. Bending energy vs. end-to-end distance x for a
DNA nanorod of length $L = 15\,\text{nm}$ (45 bp), according to
eq. (2.206), using the parameters $B = 200\,\text{pN nm}^2$,
$\tau_c = 31\,\text{pN nm}$. The energy is in units of kT.

in (2.206), or also by (2.194). The energy (2.206) is plotted in figure 2.33
for a DNA 45mer (contour length $L = 15\,\text{nm}$), using the parameters
$B = 200\,\text{pN nm}^2$, $\tau_c = 31\,\text{pN nm}$. It represents a bifurcation mathematically
similar to the Euler instability of the previous section: there are two branches
of the energy curve, and they cross, so that as the control parameter (x)
is varied, the stable (lowest-energy) solution switches from one branch to
the other (see figure 2.34, where we plot both branches of the energy
of figure 2.33 past the critical point). However, the "microscopic" physics
underlying the bifurcation is totally different in the two cases. The Euler
instability reflects a geometric nonlinearity: because of the curved geometry,
the bending energy is *linear* in the end-to-end distance, whereas the energy
of the straight rod is quadratic. In both cases, the material is within the linear
elasticity regime. In contrast, the softening transition of the DNA nanorod
is a materials nonlinearity. Nonetheless, the same mathematics describes the
"thermodynamic" (i.e., whole system) behavior of both.

From the energy (2.206) we can calculate the force $f = |\partial E/\partial x|$ that a
DNA "spring" with the ends held at a distance $x < L$ exerts on the holding
points. As we see from figure 2.35, the softening transition limits this force, in
the kinked regime, to a few pN.

It remains to connect with the experimental measurements of figure 2.25.
We write the total elastic energy of the D-DNA molecule in the form (2.171),
(2.172), with $E_d(x)$ given by (2.206). The stretching elastic energy of ss
DNA of several persistence lengths is well represented by the Marko–Siggia
expression, which for the stretching force reads

$$f_s(x) = \frac{T}{\ell_s}\left[\frac{1}{4(1 - x/L_s)^2} + \frac{x}{L_s} - \frac{1}{4}\right], \qquad (2.207)$$

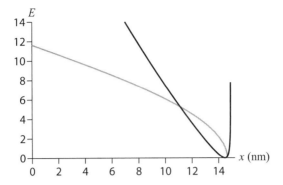

FIGURE 2.34. The two branches of the bending energy of figure 2.33 plotted past the critical point. For each value of x, the stable solution corresponds to the lower branch.

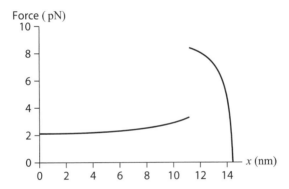

FIGURE 2.35. The force corresponding to the bending energy of figure 2.33.

where $\ell_s \approx 0.8\,\mathrm{nm}$ is the persistence length of ss DNA and $L_s = N_s \ell_s/3$ the "contour length." Integrating (2.207), the corresponding elastic energy is

$$E_s(x) = \int_0^x dx'\, f_s(x') = \frac{T}{4\ell_s} L_s \left[\frac{1}{4(1 - x/L_s)} + 2\left(\frac{x}{L_s}\right)^2 - \frac{x}{L_s} - 1 \right]. \quad (2.208)$$

Alternatively, one can use a polynomial expansion of (2.208):

$$E_s(x) = \frac{9T}{4N_s \ell_s^2} \left[x^2 + \frac{1}{N_s \ell_s} x^3 + \frac{3}{(N_s \ell_s)^2} x^4 + \cdots \right]. \quad (2.209)$$

The solid lines in figure 2.25 are calculated by the above procedure. For both cases shown ($N_d = 18\,\mathrm{bp}$ corresponding to $L = 6\,\mathrm{nm}$, and $N_d = 24\,\mathrm{bp}$

corresponding to $L = 8\,\text{nm}$), identical values of the parameters were used, namely, $\tau_c = 26.9\,\text{pN nm}$, $B = 50\,k_B T\,\text{nm}$, $\ell_s = 0.764\,\text{nm}$. It is remarkable that the softening transition of the DNA nanorod, a potentially complicated nonlinearity of the mechanics of a complex molecule, is well described by the simple model (2.206).

3

Kinematics of Enzyme Action

3.1 Introduction

Enzymes are catalysts: they speed up chemical reactions. They are also the molecular machines that generate and maintain the nonequilibrium state of the cell that is life. Chapter 4 is devoted to the molecular machine aspect, that is, the dynamics. Here we address what could be variously called the quasi-equilibrium aspects, or steady state, or kinematics, of enzyme operation, and what one learns from time-independent perturbations of this steady state. The third essential characteristic of enzymes is that they are molecules, and thus addressable by molecular control. With solid state catalysts, which are mesoscopic surfaces, the reaction speed can be modulated only by overall thermodynamic control (temperature, pressure, etc.). In the cell, there are hundreds of chemical reactions going on at any one time; the thermodynamic parameters affect the speed of all of them. Imagine we wanted to build some kind of artificial cell. It would be foolish to base it on a collection of solid state catalysts and external control of the thermodynamic parameters.

There are, roughly speaking, two main mechanisms by which enzymes provide *the* control function for the network of chemical reactions in the cell. One is simply the presence or absence of the enzyme, controlled by gene expression. The other is allosteric control: the modulation of the activity of an individual enzyme caused by binding of a specific ligand, often a small metabolite, or else caused by so-called post-translational modifications, of which the most important one is phosphorylation/dephosphorylation. These seemingly disparate mechanisms of control (ligand binding, which is non-covalent, and phosphorylation, a covalent ("chemical") modification of the enzyme) share a unifying physical basis: namely, both events cause a deformation of the enzyme molecule which results in inhibition or activation ("modulation") of the enzymatic activity.

3.2 Michaelis–Menten Kinetics

One "universal" aspect of enzyme kinetics is the dependence of the reaction speed on substrate concentration. Note that in enzymology jargon, "substrate" means reactant, and we will conform to this usage. There is an analogy here with scattering cross sections in particle physics, which can be written as the product of an interaction term and the phase space volume of the final states. The latter is the "universal" part, while the interaction term is system specific. For enzymes, one can think of certain steps the enzymatic process must follow. Substrates must bind to the enzyme. The chemical reaction must take place. Products must unbind from the enzyme. This seems obvious but in fact this subdivision of the process is somewhat arbitrary, as we can appreciate by trying to define precisely what is meant by "bind." First we give a derivation of Michaelis–Menten kinetics based on quasi-equilibrium arguments, namely, substrate binding to the enzyme is supposed in equilibrium while product formation is not. In fact, one can arrive at Michaelis–Menten kinetics under less restrictive assumptions, so the result is more general than it appears from this argument. Let us take, for simplicity, the case of one substrate, and zero concentration of products. Then as far as describing the concentration dependence of the overall speed of the process, we can choose to lump the chemical reaction step and the product unbinding step into one:

$$E + S \underset{k_{-1}}{\overset{k_1}{\rightleftharpoons}} ES \overset{k_{cat}}{\longrightarrow} E + P, \tag{3.1}$$

where E is an enzyme, S and P are substrate and product, respectively, and ES is the complex of the substrate bound to the enzyme. The k's are rates, and there is only one arrow for the second step because the product concentration $[P] = 0$ and therefore the rate for $E + P \rightarrow EP$ is zero. Now we can write the overall speed of the reaction, measured for example by the rate of disappearance of the substrate, as

$$-\frac{d[S]}{dt} = P(\text{on})[E]_{\text{tot}}k_{\text{cat}}, \tag{3.2}$$

where $P(\text{on})$ is the probability that the catalytic site on the enzyme is occupied by a substrate molecule (we assume one catalytic site per enzyme); $[E]_{\text{tot}} = [E] + [ES]$ is the total concentration of enzyme, so $P(\text{on})[E]_{\text{tot}} = [ES]$. If we assume that $P(\text{on})$ represents the equilibrium distribution (i.e., the process of the substrate binding to the enzyme is in equilibrium), then it is a Fermi–Dirac distribution because the catalytic site is either occupied by one substrate, or none:

$$P(\text{on}) = \frac{1}{e^{(\varepsilon_0 - \mu)/T} + 1}, \tag{3.3}$$

where $\varepsilon_0 < 0$ is the binding energy and μ is the chemical potential of the substrate. The concentration dependence is with μ: the "golden rule" of physical chemistry is that, in the dilute regime (see eqs. (2.15), (2.17)),

$$\mu = \mu_0 + T \ln X_S = \mu_0 + T \ln \frac{[S]}{C_w}, \tag{3.4}$$

where X_S is the mole fraction of the substrate, and $C_w \approx 55\,\text{M}$; we express concentrations [ ] in M (moles/liter). Equation (3.4) expresses the concentration dependence of the chemical potential for a dilute solution of S, which is always the case for enzymatic reactions where typical substrate concentrations are in the mM range or below. Using (3.4), (3.3), and (3.2) we obtain

$$-\frac{d[S]}{dt} = \frac{d[P]}{dt} = \frac{[E]_{\text{tot}} k_{\text{cat}}}{\frac{K_M}{[S]} + 1}, \tag{3.5}$$

where

$$K_M = C_w e^{(\varepsilon_0 - \mu_0)/T}. \tag{3.6}$$

Equation (3.5), and its generalizations, represents the (justly) celebrated Michaelis–Menten (MM) kinetics. The quantity K_M as defined by eq. (3.5) is called the Michaelis–Menten constant. For the case that substrate binding to the enzyme is in equilibrium, it is given by eq. (3.6), that is, $K_M = K_D$, the dissociation constant of the equilibrium $E + S \rightleftharpoons ES$. Integrating (3.5), $[E]_{\text{tot}}$ being of course a constant, we find the time course of the reaction,

$$S(t) - S(0) + K_M \ln \frac{S(t)}{S(0)} = -[E]_{\text{tot}} k_{\text{cat}} t, \tag{3.7}$$

where we write $S(t)$ for $[S](t)$ etc. for simplicity; $S(0)$ is the substrate concentration at time $t = 0$. We see from (3.7) that one can measure the parameters K_M and k_{cat} (and, in addition, $[E]_{\text{tot}}$) by "titration experiments" where the course of the reaction $S(t)$ is measured for different initial values of the substrate concentration, $S(0)$. Dividing eq. (3.7) by k_{cat} we see that, in fact, from this equation one determines experimentally $1/k_{\text{cat}}$, K_M/k_{cat}, and $[E]_{\text{tot}}$. One might think that the last parameter, $[E]_{\text{tot}}$, should be known by construction, but in fact that is never the case. One can, of course, measure protein concentration, but to measure the concentration of *active* enzyme (the inactive fraction being misfolded, chemically damaged, etc.) is quite another matter. For this reason, enzyme preparations are often characterized by "units of enzymatic activity" rather than so many moles of enzyme.

In practice, one often measures k_{cat} and K_M by considering simple limiting cases. For example, at short times $S(t) \approx S(0)$, and the speed of the reaction (3.5) becomes

$$-\frac{dS}{dt} = \frac{[E]_{tot} k_{cat}}{\frac{K_M}{S(0)} + 1}. \tag{3.8}$$

Varying the initial concentration of substrate, $S(0)$, one determines K_M, and varying the (nominal) concentration of enzyme one determines k_{cat}. Integrating (3.8), or else from (3.7) in the approximation $S(t) \approx S(0)$, we find

$$S(t) = S(0)\left[1 - \frac{t}{\tau}\right], \qquad \tau = \frac{K_M + S(0)}{[E]_{tot} k_{cat}}, \tag{3.9}$$

valid for short times $t \ll \tau$. Within the short time approximation, if we have lots of substrate, which means $S(0) \gg K_M$, we have

$$\tau \to \frac{S(0)}{[E]_{tot} k_{cat}} \quad \text{and} \quad S(t) = S(0) - [E]_{tot} k_{cat} t. \tag{3.10}$$

If we have little substrate ($S(0) \ll K_M$), we have

$$\tau \to \frac{K_M}{[E]_{tot} k_{cat}} \quad \text{and} \quad S(t) = S(0)\left[1 - [E]_{tot} \frac{k_{cat}}{K_M} t\right]. \tag{3.11}$$

While the reaction speed is always proportional to the total enzyme concentration, with lots of substrate the intrinsic enzymatic parameter that determines the speed is k_{cat}; with little substrate, it is k_{cat}/K_M.

On the other hand, at long times, $S(t) \ll S(0)$ and (3.7) reduces to

$$S(t) = S(0)e^{S(0)/K_M}e^{-t/\tau}, \quad t \gg \tau, \tag{3.12}$$

with τ given by (3.11).

Equation (3.5) is valid for the case of infinite dilution of the products, for instance, when the products are removed by another enzymatic reaction. In general, we must consider binding of the products to the enzyme: if a product is occupying the catalytic site, a new substrate cannot come in. To take product inhibition into account, we add one step to the reaction scheme (3.1):

$$E + S \underset{k_{-1}}{\overset{k_1}{\rightleftharpoons}} ES \overset{k_{cat}}{\longrightarrow} EP \rightleftharpoons E + P. \tag{3.13}$$

Now, there are three possible "states" of the enzyme: E, ES, and EP, and the partition sum expressing these different states is

$$Z = 1 + e^{\mu_S/T}e^{-\varepsilon_S/T} + e^{\mu_P/T}e^{-\varepsilon_P/T}, \tag{3.14}$$

written in the grand canonical ensemble, allowing at most one particle of each species (S or P) to bind to the enzyme. We want the concentration dependence so we exhibit the dilution term in the chemical potential (see (3.4)):

$$\mu_S = \mu_S^0 + T \ln \frac{[S]}{C_w}, \qquad \mu_P = \mu_P^0 + T \ln \frac{[P]}{C_w}. \tag{3.15}$$

The probability of the state ES is

$$P(\text{on}) = \frac{e^{(\mu_S - \varepsilon_S)/T}}{Z} = \frac{1}{1 + \frac{K_S}{[S]}\left(1 + \frac{[P]}{K_P}\right)}, \tag{3.16}$$

where we have used (3.14) and (3.15) and defined

$$K_S = C_w e^{(\varepsilon_S - \mu_S^0)/T}, \qquad K_P = C_w e^{(\varepsilon_P - \mu_P^0)/T}. \tag{3.17}$$

From (3.13), the rate at which the substrate is consumed is

$$-\frac{d[S]}{dt} = [E]_{\text{tot}} P(\text{on}) k_{\text{cat}} = \frac{[E]_{\text{tot}} k_{\text{cat}}}{1 + \frac{K_S}{[S]}\left(1 + \frac{[P]}{K_P}\right)}. \tag{3.18}$$

In the steady state of (3.13), by which we mean $[EP] = $ const. over several enzymatic cycles, this is also the rate at which the product appears, that is, $-d[S]/dt = d[P]/dt$. There is no difficulty extending this calculation to the case of two or more products. For example, for the situation

$$E + S \rightleftharpoons ES \xrightarrow{k_{\text{cat}}} EPQ \rightleftharpoons E + P + Q \tag{3.19}$$

(a substrate molecule S gets chopped up into two pieces P and Q) we find

$$-\frac{d[S]}{dt} = \frac{[E]_{\text{tot}} k_{\text{cat}}}{1 + \frac{K_S}{[S]}\left(1 + \frac{[P]}{K_P} + \frac{[Q]}{K_Q} + \frac{[P][Q]}{K_{PQ}}\right)}, \tag{3.20}$$

the five terms in the denominator reflecting the terms in the partition sum corresponding to the states E, ES, EP, EQ, EPQ. Often the product term $[P][Q]$ can be neglected.

Similarly, let us extend (3.5) to the case of two substrates. The reaction schematics is

$$E + A + B \rightleftharpoons \begin{Bmatrix} EA + B \\ EB + A \end{Bmatrix} \rightleftharpoons EAB \xrightarrow{k_{\text{cat}}} E + P. \tag{3.21}$$

There are four states: E, EA, EB, EAB; the probability that the enzymatic site is occupied by A only is $P(A)$ and so on. The grand partition sum is

$$Z = 1 + e^{(\mu_A - \varepsilon_A)/T} + e^{(\mu_B - \varepsilon_B)/T} + e^{\mu_A/T} e^{\mu_B/T} e^{-(\varepsilon_A + \varepsilon_B)/T}, \tag{3.22}$$

where we have assumed, for simplicity, that the binding energy ε_{AB} for *both* A and B bound to the enzyme is the sum $\varepsilon_{AB} = \varepsilon_A + \varepsilon_B$. This is generally not the case due to "cooperative effects," that is, the interaction between the two bound substrates, which can be direct (e.g., electrostatic) or mediated by the enzyme's strain field. Then there is one more parameter in the model, namely ε_{AB}.

The speed of the process (3.21) is (in the steady state, which means that the concentrations EA, EB, EAB do not change over several enzymatic cycles)

$$\frac{d[P]}{dt} = P(AB)[E]_{\text{tot}}k_{\text{cat}}. \tag{3.23}$$

From (3.22),

$$P(AB) = \frac{1}{Z}e^{(\mu_A + \mu_B - \varepsilon_A - \varepsilon_B)/T}, \tag{3.24}$$

and writing the chemical potentials as in (3.15) we find

$$P(AB) = \frac{1}{1 + \frac{K_A}{[A]} + \frac{K_B}{[B]} + \frac{K_A K_B}{[A][B]}} = \frac{1}{\left(1 + \frac{K_A}{[A]}\right)\left(1 + \frac{K_B}{[B]}\right)}, \tag{3.25}$$

where $K_A = C_w e^{(\varepsilon_A - \mu_A^0)/T}$ etc. Clearly, with the same method that led to (3.20) and (3.25), we can deal with any number of substrates and products. Perhaps the most typical situation is two substrates and two products:

$$E + A + B \longrightarrow E + C + D \tag{3.26}$$

with the catalytic step

$$EAB \xrightarrow{k_{\text{cat}}} ECD. \tag{3.27}$$

Then,

$$\frac{d}{dt}[C] = \frac{d}{dt}[D] = P(EAB)[E]_{\text{tot}}k_{\text{cat}}, \tag{3.28}$$

with

$$P(EAB) = \left\{1 + \frac{K_A}{[A]} + \frac{K_B}{[B]} + \frac{K_A K_B}{[A][B]}\left(\frac{[C]}{K_C} + \frac{[D]}{K_D} + \frac{[C][D]}{K_C K_D}\right)\right\}^{-1}, \tag{3.29}$$

where we have assumed that binding energies sum ($\varepsilon_{AB} = \varepsilon_A + \varepsilon_B$ etc.) and we have considered the states E, EA, EB, EAB; EC, ED, ECD. There could be more states, say EAC or EBD; then there are correspondingly more terms in (3.29).

Two further comments are in order, which we refer to the simplest case (3.13). In a closed system, $[S] + [P] = $ const. For example, suppose we start at time zero with $S(0) = S_0$, $P(0) = 0$ (we drop the square brackets). Then (3.18) becomes

$$-\frac{dS}{dt} = \frac{[E]_{\text{tot}}k_{\text{cat}}}{1 + \frac{K_S}{S(t)}\left(1 + \frac{S_0 - S(t)}{K_P}\right)}, \tag{3.30}$$

which we can of course integrate to obtain $S(t)$. The other remark is that in general we also have to consider the reverse reaction, that is,

$$E + S \rightleftharpoons ES \underset{\tilde{k}_{cat}}{\overset{k_{cat}}{\rightleftharpoons}} EP \rightleftharpoons E + P. \tag{3.31}$$

In this case we write the dynamics as

$$\frac{d[S]}{dt} = -P(ES)[E]_{tot}k_{cat} + P(EP)[E]_{tot}\tilde{k}_{cat} \tag{3.32}$$

and

$$P(ES) = \left\{ 1 + \frac{K_S}{[S]}\left(1 + \frac{[P]}{K_P}\right)\right\}^{-1},$$

$$P(EP) = \left\{ 1 + \frac{K_P}{[P]}\left(1 + \frac{[S]}{K_S}\right)\right\}^{-1}. \tag{3.33}$$

We now come back to the beginning of our discussion, with eqs. (3.1) and (3.5). A slightly different way to think about the process (3.1) is through the individual rates k_1, k_{-1}, k_{cat}. We write the rate equations:

$$\frac{d[S]}{dt} = -k_1[E][S] + k_{-1}[ES], \tag{3.34}$$

$$\frac{d}{dt}[ES] = -(k_{-1} + k_{cat})[ES] + k_1[E][S]. \tag{3.35}$$

We assume a steady (or quasi-equilibrium) state, meaning

$$\frac{d}{dt}[ES] = 0 \quad \Rightarrow \quad [E][S] = \frac{k_{-1} + k_{cat}}{k_1}[ES]. \tag{3.36}$$

Using (3.36) in (3.34) we find

$$\frac{d[S]}{dt} = -k_{cat}[ES], \tag{3.37}$$

which is the same as (3.2): in quasi-equilibrium as defined above, the reaction speed is k_{cat} times the concentration of enzyme with a substrate bound. Now we want to write $[ES]$ in terms of $[E]_{tot}$ and $[S]$: since

$$[E] + [ES] = [E]_{tot} \tag{3.38}$$

we have, using (3.36),

$$[ES] = [E]_{tot} - [E] = [E]_{tot} - \frac{k_{-1} + k_{cat}}{k_1}\frac{1}{[S]}[ES] \Rightarrow [ES] = \frac{[E]_{tot}}{1 + \frac{k_{-1} + k_{cat}}{k_1}\frac{1}{[S]}}, \tag{3.39}$$

and therefore,

$$-\frac{d[S]}{dt} = \frac{[E]_{\text{tot}}k_{\text{cat}}}{1 + \dfrac{k_{-1} + k_{\text{cat}}}{k_1}\dfrac{1}{[S]}}, \tag{3.40}$$

which is the same as (3.5), with the Michaelis–Menten constant given by

$$K_{\text{M}} = \frac{k_{-1} + k_{\text{cat}}}{k_1}. \tag{3.41}$$

The purpose of this exercise is to see that expression (3.5) is more general than the interpretation (3.6) for K_{M}. In general, K_{M} is not the same as the dissociation constant (3.6). If $k_{-1} \gg k_{\text{cat}}$ then $K_{\text{M}} \approx k_{-1}/k_1 = K_{\text{D}}$, consistent with (3.6); otherwise K_{M} is given by (3.41), which says that there are two channels through which ES can disappear: the "dissociation channel" with rate k_{-1} and the "chemical reaction channel" with rate k_{cat}.

One interesting consequence of (3.41) is the following: If we measure K_{M} and k_{cat}, by substrate titration experiments, and $K_{\text{D}} = k_{-1}/k_1$ (the latter is tricky for a one-substrate enzyme, but not too difficult to measure for a two-substrate enzyme), we can determine the *rates* k_1 and k_{-1}:

$$k_1 = \frac{k_{\text{cat}}}{K_{\text{M}} - K_{\text{D}}}, \qquad k_{-1} = \frac{K_{\text{D}}}{K_{\text{M}} - K_{\text{D}}}k_{\text{cat}}. \tag{3.42}$$

Finally, note that the process of product release $EP \to E + P$ would seem, conceptually, an ideal experimental realization of the Kramers problem of escape over a barrier, which we saw in chapter 1: the product P is created at a definite time in the potential well which is the enzyme's active site. However, what distinguishes enzymes from solid state catalysts is the coupling of the catalytic process to conformational motion. The enzyme deforms as it binds the substrates, and deforms again as it releases the products. In general, the enzyme goes through a chemo-dynamic cycle, which we can imagine represented in the stress–strain plane (think of the representation of the Carnot cycle in the P–V plane, though, I hasten to add, enzymes are *not* thermodynamic engines: they are chemo-dynamic engines). If the cycle has nonzero area, it can perform work. This work may increase the free energy of a different system that the cell cares to keep out of equilibrium. For instance, ionic pumps (e.g., the sodium–potassium pump) are enzymes that, through a series of ligand-induced conformational transitions, pump specific ions in or out of the cell *against a concentration gradient*. The reaction catalyzed by the enzyme—ATP hydrolysis—has nothing to do with those ions. Chemo-mechanical coupling in enzymes is the molecular-scale mechanism that maintains the nonequilibrium state of the cell; for example, it maintains ionic gradients across the cell membrane.

To be specific, let us think of the chemo-dynamic cycle in the stress–strain plane of the enzyme. In the limiting case that the area of the cycle is zero, no work is performed. This is the case if there is no conformational change: the case of the solid state catalyst. What would happen if all enzymes in the cell were like solid state catalysts? All reactions would quickly be driven to equilibrium and the cell would die. A cell in equilibrium is a dead cell. We see that chemo-mechanical coupling in enzymes is the molecular basis of life. Another way that the area of the cycle can be zero is that the enzyme moves in the stress–strain plane back and forth along a line. This means that the forward and backward conformational motions are the time reverse of each other. Then, also, no work can be performed. Figure 3.1 shows a hypothetical chemo-dynamic cycle in the stress–strain (i.e., σ–u) plane.

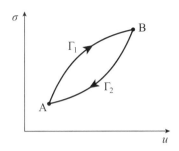

FIGURE 3.1. Representation of a chemo-dynamic cycle in the stress–strain plane (σ–u plane).

The process Γ_1 corresponds to binding the substrates; Γ_2 corresponds to releasing the products. The chemical reaction happens at B. The area inside the loop (times the volume of the enzyme) is the work performed. Michaelis–Menten kinetics makes, of course, no reference to the chemo-dynamic cycle of the figure; the parameters k_{cat}, K_M are not related in any simple way to the trajectories. Nonetheless, the Michaelis–Menten description provides a way to measure some parameters that are intrinsic to the enzyme. An analogy is obtaining the scattering matrix from measurements of the scattering cross section. One reasonable—and, it turns out, feasible—experimental approach to investigating the chemo-dynamic cycle is to mechanically stress the enzyme and observe the effect on the intrinsic enzymatic parameters. This is very basic. To do an experiment, we need to identify a field that couples to the process or phenomenon we want to investigate. For example, to investigate ferromagnetism we need experiments with a controlled magnetic field. Once we know that the enzyme deforms during the chemo-dynamic cycle, we want experiments where we apply controlled stresses. Stress is the field that couples to deformation.

3.3 The Method of the DNA Springs

The DNA nanorod discussed in chapter 2 (a ds DNA molecule of length somewhere between ~ 2 and ~ 6 helix turns) can be used as a leaf spring

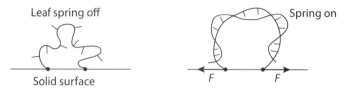

FIGURE 3.2. Sketch of the DNA spring as an addressable leaf spring. In the ds form, the spring exerts a stress on the solid surface.

to exert controlled mechanical stresses at the molecular scale. From the discussion in chapter 2, the elastic energies and forces resulting from bending the DNA nanorod are of order $\sim 10\,kT$ and a few pN, respectively. These are also the forces and mechanical work that will significantly deform—but not destroy—the folded state of an enzyme. Of course, this is not coincidental: ultimately the same forces between groups of atoms hold together the structure of both the enzyme and the DNA nanorod: hydrogen bonds, hydrophobic contacts, etc. As a mechanical device, the DNA nanorod is addressable: the bending stiffness is zero in the ss form, substantial in the ds form. The leaf spring can be "turned on" by a chemical signal: the oligonucleotide with the specific complementary sequence (figure 3.2). It can be turned off by another chemical signal: a competitor DNA strand, an enzyme that cuts the DNA, etc. In summary, if one has a way to attach the ends of the nanorod to two points on a solid surface a few nm apart, then one can exert a controlled force on these attachment points (tending to separate them) of a few pN.

It is possible to attach the ends of the nanorod to the surface of an enzyme, or more generally a protein. One engineers two "chemical handles" on the surface of the protein, for example by introducing, through mutagenesis, two Cys residues. Cystein has a uniquely reactive S–H (thiol) group. It is often possible to substitute a pair of amino acids at the surface of the molecule with Cys, without damaging the enzyme, as long as one works away from the active site. To attach the end of a single DNA strand to the Cys, one starts with a synthetic oligo end-functionalized with a primary amino group. This is attached to the Cys through a cross-linker (an organic molecule that reacts on one side with the S–H group of the Cys, and on the other side with the NH_2 group on the DNA). The reason not to use direct attachment of thiol-functionalized DNA to the Cys is in order to avoid, during the synthesis, the competing reactions of disulfide bond formation between two enzyme molecules or two DNA molecules. The enzyme–DNA chimera is constructed sequentially in multiple steps A → B → C → D (figure 3.3): first one DNA "arm" is attached, then the other, then the two are ligated. If one, instead, tries to attach the whole DNA (ab) functionalized at both ends in one step,

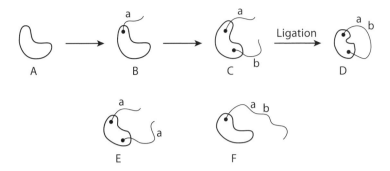

FIGURE 3.3. Scheme of the step-by-step construction of the enzyme–DNA chimeras.

one ends up with a mixture containing a large fraction of the species F, which is difficult to separate from the correct construct D, since they are chemically almost identical molecules. In the sequential construction, one also produces unwanted species, for example the species E along with B in the first step. However, it is easy to separate E from B, in this case using ion exchange chromatography, since the two molecules have very different charge. In the end, one can synthesize enzyme–DNA chimeras such as the example represented in figure 3.4. When making a mental image of these structures, one has to abstract from the static cartoon of the figure and remember that thermal fluctuations allow the structure in equilibrium to explore a range of quite different conformations. For example, the cross-linkers allow rotational freedom so that the enzyme is in constant rotational diffusion with respect to the plane of the DNA spring. On the timescales of enzyme action (ms), the DNA spring seen by the enzyme is actually a diffuse closed shell, not a ribbon in any fixed position. To get a sense of the timescales, consider the rotational diffusion of an object the size of the enzyme. If $c(\theta, t)$ is the concentration of molecules with a particular orientation θ at time t (or, if preferred, the probability that a molecule has orientation θ at time t), we may write the diffusion equation

$$\frac{\partial c(\theta, t)}{\partial t} - D_{\text{rot}} \frac{\partial^2 c}{\partial \theta^2} = 0. \tag{3.43}$$

The rotational diffusion coefficient, as we see from (3.43), has dimensions $[D_{\text{rot}}] = 1/\text{time}$, and we have written (3.43) in 1-D for simplicity (i.e., it describes rotational diffusion around a fixed axis). In analogy with translational diffusion (chapter 1), we may introduce a rotational mobility μ_{rot} through the relation

$$\langle \dot{\theta} \rangle = \mu_{\text{rot}} \tau, \tag{3.44}$$

FIGURE 3.4. Cartoon of the RLuc enzyme–DNA construction. The renilla luciferase is the PDB structure 2PSJ. The DNA is from the nucleosome structure PDB 1KX5. Adapted from Tseng and Zocchi (2013).

where τ is the applied torque and the brackets signify ensemble average. We get μ_{rot} from hydrodynamics: for a sphere of radius R in a fluid of viscosity η it is

$$\mu_{\text{rot}} = \frac{1}{8\pi\eta R^3},\tag{3.45}$$

while mobility and the diffusion coefficient are connected by the Einstein relation

$$D_{\text{rot}} = T\mu_{\text{rot}},\tag{3.46}$$

and therefore, numerically,

$$D_{\text{rot}} = \frac{kT}{8\pi\eta R^3}\tag{3.47}$$

for the sphere. The solutions of (3.43) are such that (see chapter 1)

$$\langle \theta^2 \rangle = 2D_{\text{rot}}t,\tag{3.48}$$

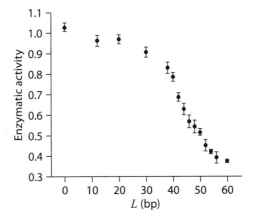

FIGURE 3.5. Measured activity modulation for the enzyme guanylate kinase under increasing mechanical stress. The enzyme is coupled to a 60 mer DNA spring. With the spring in the ss form ($L = 0$ point) there is no stress on the enzyme. For increasing length L of the complementary strand hybridized to the DNA spring, the mechanical stress on the enzyme increases and the speed of the enzymatic reaction decreases. Adapted from Choi and Zocchi (2007).

so the diffusion time over, say, 1 radian is

$$\tau_{\mathrm{D}} = \frac{1}{2D_{\mathrm{rot}}} = \frac{4\pi\eta R^3}{kT}. \tag{3.49}$$

For a 4 nm-diameter sphere in water we find $\tau_{\mathrm{D}} \approx 20\,\mathrm{ns}$. Thus the rotational diffusion of the enzyme around the axis provided by the DNA spring is very fast compared to the timescales of the enzymatic cycle. The relative rotational diffusion of the enzyme and the DNA spring could be measured, for example, by fluorescence correlation spectroscopy.

Mechanically stressing an enzyme is a general way to modulate its enzymatic activity. "General" means not enzyme specific. In contrast, inhibitors—small molecules that bind to, and block, the catalytic site—are of course enzyme specific: the inhibitor of one enzyme will not inhibit another enzyme. Figure 3.5 shows the progressive inhibition of the enzyme guanylate kinase (GK) as the mechanical stress on the enzyme is increased, by means of an increasingly stiff DNA spring. The DNA strand covalently attached to the enzyme is 60 bases long, and the stiffness modulation is obtained by hybridization with complementary strands of different lengths (figure 3.6). These enzyme–DNA chimeras allow precise measurements of the effect of mechanical stress on the enzymatic activity because one can compare

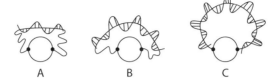

FIGURE 3.6. The sketch represents a molecular construction consisting of an enzyme and a 60-bases-long DNA spring. In species A, the DNA spring is hybridized with a 20 mer complementary. There is no stress on the enzyme because the ss parts of the DNA (two segments each 20 bases long) provide enough slack so that the ds part does not have to bend. In species B, the DNA spring is hybridized to a 40 mer complementary. The ds part of the DNA has to bend and the two ss segments (each 10 bases long) have to stretch, so the enzyme is under mechanical stress. In species C, the DNA spring is fully hybridized (to the complementary 60 mer) and the enzyme is under a bigger stress than species B. Compare with figure 3.5.

samples where the amount of enzyme is by construction the same, but the mechanical stress is turned on, or not, depending on the complementary oligomer added into the different samples. Moreover, one can measure by standard biochemical methods the intrinsic parameters of the enzyme, such as Michaelis–Menten constants and k_{cat}, in the presence and absence of mechanical stress.

The detailed studies performed on the enzyme guanylate kinase form one basis for the continuum mechanics approach to enzyme dynamics adopted in the next chapter. GK catalyzes the reaction of transferring a phosphate group from ATP to GMP:

$$GMP + ATP \xrightarrow{GK} GDP + ADP, \tag{3.50}$$

that is, the phosphorylation of GMP. It is a monomeric enzyme about 4 nm in size (about 200 amino acids), with a characteristic kidney-bean shape (see figure 4.7). The binding sites for the substrates lie on the concave surface in between the two lobes of the structure. Binding of GMP elicits a large conformational change of the enzyme, the two lobes closing upon each other (see figure 4.7) through an ~ 1 nm-size motion. This enzyme represents a classic example of the induced-fit mechanism discovered by Koshland in the middle of the twentieth century. The catalytic site is active in the "closed" conformation of the enzyme, that is, only when GMP is bound. The closed conformation also excludes unstructured water from the reaction center, thus avoiding the competing reaction of ATP hydrolysis. It simply wouldn't do for the cell to hydrolyze ATP for no purpose. The precise values of the intrinsic

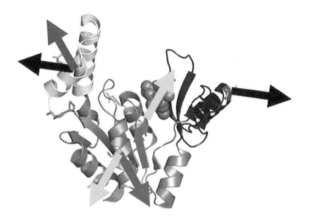

FIGURE 3.7. Different application points of the stress
(arrows) for the enzyme guanylate kinase. They are
realized by attaching the DNA spring to the corresponding
locations on the surface of the enzyme.

parameters of GK vary of course, depending on the species, but roughly
speaking the catalytic rate is in the range $k_{cat} \sim 100–400$ Hz. For GK from
Mycobacterium tuberculosis (the model system for some of the mechanical
experiments described below) the Michaelis–Menten constant for ATP is
$K_A \sim 2$ mM (to be compared with typical ATP concentrations inside the cell
of ~ 5 mM), while the Michaelis–Menten constant for GMP is $K_G \sim 200\,\mu$M.
Let us now examine the effects of mechanical stress on this enzyme. If the
force is applied as shown by the black arrows in figure 3.7 (this is achieved
by attaching the DNA spring to the corresponding locations on the surface
of the enzyme, defined chemically by Cys residues substituted by mutagenesis), the reaction slows down, as shown in figure 3.5. More specifically,
GMP binding is impaired while ATP binding and the catalytic rate k_{cat} are
essentially unaffected. For example, under the conditions corresponding to
the titration experiments of figure 3.8 (stress according to the black arrows
in figure 3.7), the Michaelis–Menten constant for GMP, K_G, is increased by a
factor 10 under stress (recall that Michaelis–Menten constants are essentially
dissociation constants: a larger value means weaker binding), while K_A and
k_{cat} change less than 30%. If it is justified to view the enzyme as a mechanical
system, and the DNA spring as an external force acting on it, then the fact
that the mechanical stress according to the black arrows (figure 3.7) reduces
the binding affinity for GMP follows from general thermodynamic principles.
Recall the Le Chatelier principle:

> If the external conditions of a thermodynamic system are altered, the
> equilibrium of the system will tend to move in such a direction as to
> oppose the change in the external conditions.

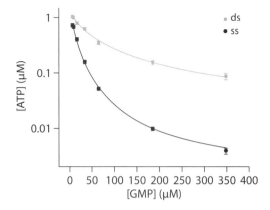

FIGURE 3.8. GMP titration experiments for the enzyme guanylate kinase under mechanical stress ("ds": squares) and without stress ("ss": circles). Plotted is the ATP concentration remaining in the reaction mixture after a given time. Since the reaction consumes ATP, the enzyme under stress is slower. Adapted from Choi et al. (2005).

In our case, GMP binding drives the open-to-closed conformational change of the enzyme; this means that the free energy of the GMP–GK bound system is lower in the closed conformation compared to the open conformation, that is, GMP binds stronger to the closed conformation than the open one. A force opposed to the open-to-closed conformational change increases the free energy of the closed conformation, compared to the open one; according to Le Chatelier, this force must therefore have the effect of shifting the equilibrium towards the open conformation, that is, it must cause a lowering of the binding affinity (increase of the dissociation constant) for GMP.

If the force is applied as shown by the dark gray arrows in figure 3.7, then also the reaction slows down, but for a different reason. Namely, the catalytic rate k_{cat} is reduced, whereas K_G and K_A are essentially unaffected. As an example, under the conditions of the titration experiments of figure 3.9, k_{cat} is reduced, under stress, by a factor 0.4 while K_A and K_G are unaffected. Under these same conditions but with the DNA spring attached according to the black arrows, K_G is increased, under stress, by a factor 2.8 while k_{cat} and K_A are unaffected. In summary, forces applied at different locations on the enzyme elicit different responses, not different magnitudes of the same response. These experiments show that there are many different conformational states that the enzyme can adopt, and that these different states or deformations of the structure are accessible by perturbing the enzyme mechanically. It would be a mistake to conclude from the X-ray structures, such as figure 4.7, that the transition from the open to the closed

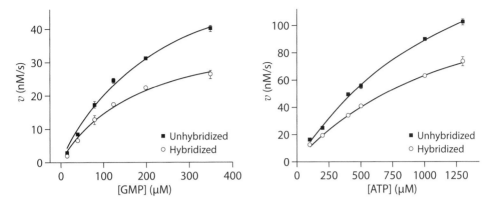

FIGURE 3.9. GMP and ATP titration experiments for the enzyme GK under mechanical stress (circles) and without stress (squares). The stress is exerted according to the dark gray arrows in figure 3.7. The quantity plotted is the initial speed of the enzymatic reaction. Adapted from Tseng, Wang, and Zocchi (2010).

conformation is the only deformation possible for the folded structure of the enzyme. On the contrary, many different deformed states are possible, accessible by mechanical perturbation. GMP binding elicits the specific deformation shown in figure 4.7, but a different mechanical perturbation will elicit a different deformation. We will see in chapter 4 a quite different set of experiments, based on the peculiar kinetics of ligand binding to myoglobin at low temperature, which highlight that many different conformational states (i.e., many different deformed states) are available to the folded structure of a protein. The method of the DNA springs provides a means of accessing some of these different conformations within an equilibrium experiment. We will see in chapter 4 that nano-rheology provides the means of accessing different deformed states dynamically, that is, within a nonequilibrium experiment.

3.4 Force and Elastic Energy in the Enzyme—DNA Chimeras

Now we examine what the forces that elicit a sizeable modulation of enzymatic activity are, and what the elastic energy injected into the enzyme structure by these forces is. If we take the overall speed of the reaction as a measure of enzymatic activity, then a model experimental system for quantitative analysis is provided by the RLuc chimeras. These molecular constructs consist of the enzyme renilla luciferase coupled to a DNA spring (figure 3.4). The luciferase is from *Renilla reniformis*, which is a sea colony of the class Anthozoa, the same class as sea anemones. This enzyme catalyzes a luminescent reaction, namely the oxidation, by molecular oxygen dissolved in the water, of its substrate coelenterazine. With a low quantum yield

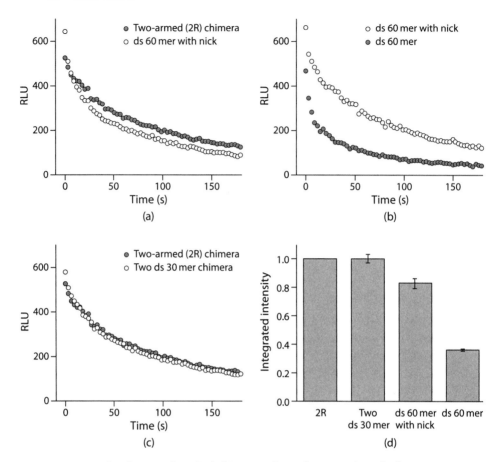

FIGURE 3.10. Luminescence intensity (arbitrary units) over the course of time for the RLuc chimera under different states of mechanical stress. Adapted from Tseng and Zocchi (2013).

of ∼ 5%, the reaction produces a blue photon ($\lambda \approx 470\,\text{nm}$), and for this reason, precise measurements of the enzymatic activity are relatively easy, based on the emitted light. Figure 3.10 shows the time course of the reaction's luminescence for different states of mechanical stress of the enzyme. The DNA spring is 60 bases long, and different configurations are used: ss, ds with a nick, ds without nick. Recall that, for quantitative measurements with enzymes, one has to compare samples that contain, by construction, the same amount of enzyme. For this reason, the reaction speed with the DNA spring in the ds nicked form is compared with the speed for the unligated, unhybridized, "two-arms" chimera. The latter is species C in figure 3.3, and the former is obtained by hybridization of this species with the complementary DNA strand. A zero-force control is provided by hybridizing the same species C with two separate 30 mers, one complementary to strand a, the other to

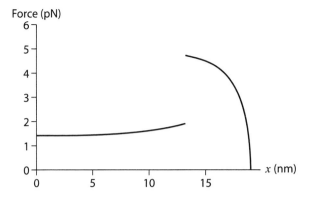

FIGURE 3.11. Calculated force vs. end-to-end distance x for the 60 mer nicked DNA spring. The graph is a plot of eq. (3.52) using $L = 60 \times 0.33 = 19.8$ nm, $\tau_c = 27$ pN nm, $B = 200$ pN nm^2.

strand b. This control reassures us that the rather small modulation of activity observed with the nicked 60 mer spring is indeed significant. Similarly, the speed for the ds non-nicked form is compared to the ds nicked form, both forms originating from species D in figure 3.3, by hybridization with one complementary 60 mer and two separate 30 mers, respectively. For the case of the nicked spring, we can calculate the force exerted on the enzyme exactly, using the results of chapter 2. Let us first suppose that the enzyme is not deformed; then we know the end-to-end distance s of the DNA spring from the geometry of the molecular construct in figure 3.4:

$$s = (\text{distance between Cys residues}) + 2$$
$$\times (\text{length of enzyme–DNA cross-linker}),$$

which gives $s = (1.9 + 2 \times 2.1)$ nm $= 6.1$ nm.

The force f is obtained from the formula for the elastic energy vs. end-to-end distance x for the DNA spring (chapter 2):

$$E(x) = \begin{cases} -10\dfrac{B}{L^2}(x - x_0) - T\ln\left(\dfrac{L - x}{L - x_0}\right) & \text{for } x \geq x_c, \\[2mm] \tau_c \arccos\left(\dfrac{x}{2R}\right) & \text{for } x < x_c. \end{cases} \tag{3.51}$$

Here, L is the contour length of the DNA, τ_c the critical bending torque, B the bending modulus, $x_0 = L[1 - TL/(10B)]$, $R = (L/2)(1 - \gamma^2/90)$, $\gamma = L\tau_c/(2B)$, and x_c is obtained by equating the upper and lower expressions in (3.51).

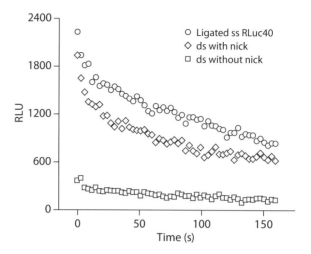

FIGURE 3.12. Measured modulation of enzymatic activity for the RLuc chimera with a 40 mer DNA spring. Adapted from Tseng and Zocchi (2014).

The force

$$
f(x) = \left| \frac{\partial E}{\partial x} \right| = \begin{cases} \dfrac{10B}{L^2} - \dfrac{T}{L-x} & \text{for } x \geq x_c, \\[2ex] \dfrac{\tau_c}{2R} \dfrac{1}{\sqrt{1-(x/2R)^2}} & \text{for } x < x_c, \end{cases}
\tag{3.52}
$$

is plotted in figure 3.11 using the parameter values for the 60 mer nicked spring ($L = 19.8\,\text{nm}$, $\tau_c = 27\,\text{pN nm}$, $B = 200\,\text{pN nm}^2$). We see from the figure that for $x = s = 6.1\,\text{nm}$, the DNA spring is in the kinked regime, and that the force is

$$
f = f(s) = 1.5\,\text{pN}.
\tag{3.53}
$$

We also see that we can relax the assumption that the enzyme is not deformed: unless the enzyme's deformation is several nm (which is unlikely without complete loss of activity), the force is still essentially the same. In conclusion, for this particular case, a force $f = 1.5\,\text{pN}$ slows down the reaction by a factor 0.8 (figure 3.10).

Figure 3.12 shows another set of measurements for the same system, except that the DNA spring is 40 bases long. There is a bigger effect on the enzyme, compared to the 60 mer spring (figure 3.13). In fact, the factor of 0.15 for the reduction in activity with the non-nicked spring should be taken as an upper limit for that situation, as the yield of correctly ligated chimeras in the samples (i.e., the yield of species D in figure 3.3) must be < 1.

The force $f(x)$, calculated from (3.52) using the parameter values for the 40 mer nicked spring ($L = 40 \times 0.33 = 13.2\,\text{nm}$, $\tau_c = 27\,\text{pN nm}$, $B = 200\,\text{pN nm}^2$) is plotted in figure 3.14. For $s = 6.1\,\text{nm}$, we see that

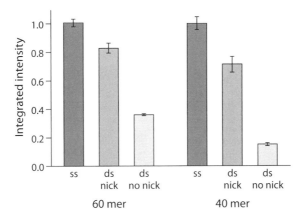

FIGURE 3.13. Comparison of the modulation of enzymatic activity obtained with the 60 mer and the 40 mer DNA springs. The areas under the corresponding curves of figures 3.10 and 3.12 are plotted, normalized by the zero-stress case. Adapted from Tseng and Zocchi (2014).

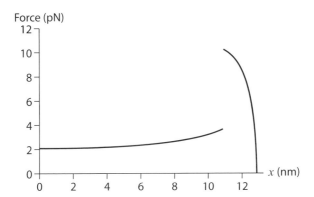

FIGURE 3.14. Calculated force vs. end-to-end distance x for the 40 mer nicked DNA spring. The graph is a plot of eq. (3.52) using $L = 13.2$ nm, $\tau_c = 27$ pN nm, $B = 200$ pN nm^2.

$f(s) = 2.4$ pN. Thus a force $f = 2.4$ pN slows down the reaction by a factor 0.7. For the non-nicked case, one cannot at present give a precise value of the force, for two reasons. First, the critical bending torque τ_c has not been measured directly for this situation. Further, fraying of the DNA at the ends may contribute to limit the force in this case. Nonetheless, the indications are that the effective value of τ_c to be used for the purpose of calculating the force in the non-nicked case is not *very* different from the nicked case (by which we mean, less than a factor 2 different). Support for this statement comes from indirect measurements of the elastic energy in the non-nicked case, based on the thermal melting transition of D-shaped DNA molecules. They find $\tau_c \approx 31\,\mathrm{pN\,nm}$. Further, the value $\tau_c = 32\,\mathrm{pN\,nm}$ would be consistent with the onset of departure from worm-like-chain behavior in the original cyclization experiments of Cloutier and Widom. Finally, from figure 3.13 we see that the nicked 40 mer spring has almost half the effect on the enzyme of the intact 60 mer spring; thus the presence of a nick does not soften the DNA spring dramatically. Summarizing the various clues, we feel that the value for τ_c for non-nicked DNA lies probably somewhere in the range 31–36 pN nm.

There is a specialized class of enzymes—motor proteins—for which detailed measurements of the modulation of the enzymatic cycle caused by an external force can be performed. The experiments of Block and coworkers in the 1990s established force–velocity curves for the motion of the enzyme kinesin along the microtubule. These are single molecule experiments where a kinesin molecule "carries" a micron-size bead while walking along the microtubule. The position of the bead over the course of time is measured by optical trapping interferometry, while the optical trap also provides a calibrated load force contrasting the motion. Several variants of this experiment have been performed, notably one version where the optical trap is moved by a feedback mechanism in order to provide, in effect, a force clamp (figure 3.15).

Kinesin steps along the microtubule through a cycle of conformational changes involving two "legs" (actually called "heads"), hydrolyzing one ATP molecule per step. With the enzyme under mechanical stress, the speed of the cycle is reduced and indeed vanishes for a "stall force" of about 7 pN (figure 3.16). The force–velocity curve is roughly (but not exactly) linear, a force of $\sim 4\,\mathrm{pN}$ reducing the speed by a factor ~ 0.5. As a function of ATP concentration, the velocity V of kinesin follows Michaelis–Menten kinetics:

$$V = \frac{V_{\max}}{1 + K_A/[\mathrm{ATP}]},\qquad(3.54)$$

so in these same experiments one can measure an effective Michaelis–Menten constant K_A, defined by (3.54). Under mechanical stress, K_A increases from $90\,\mu\mathrm{M}$ with a load $f = 1\,\mathrm{pN}$, to $140\,\mu\mathrm{M}$ for $f = 3.6\,\mathrm{pN}$, to $310\,\mu\mathrm{M}$ for $f = 5.6\,\mathrm{pN}$. The mechano-chemical cycle of this motor involves, besides ATP hydrolysis (i.e., binding ATP, hydrolyzing ATP, releasing ADP and phosphate), also the binding–unbinding of the two heads, alternately, to the microtubule. In the mechanical experiments, the force is applied between the microtubule and the body of the motor, so the stress on the enzyme is across the body and, alternately, one or the other head. Thus the geometry is different from the enzyme–DNA chimeras, but nonetheless, in all these experiments, similar forces elicit similar modulations of the speed of the enzymatic cycle.

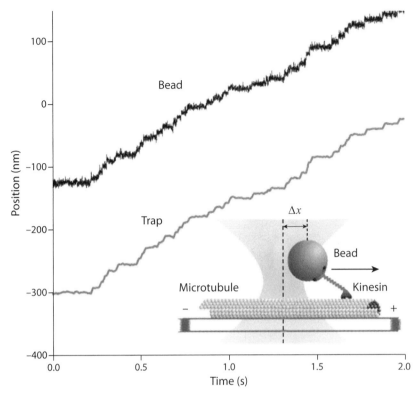

FIGURE 3.15. Schematic of the force clamp used for the single molecule experiments on kinesin. The body of the motor is attached to a micron-size bead which is held in the optical trap. A feedback mechanism moves the trap, maintaining constant force on the bead in the opposite direction to the motion. The traces show the bead's movement and the corresponding displacement of the trap. Adapted from Visscher, Schnitzer, and Block (1999).

The enzyme–DNA chimeras (figure 3.4) are molecular constructions with built-in stress. The corresponding positive elastic energy of these molecules can be measured for the case of the nicked spring, using the method described in chapter 2 for D-shaped DNA. Namely, two chimera molecules can form a dimer (figure 3.17), which releases the stress. The elastic energy of the monomer is measured from the equilibrium concentrations of monomers and dimers:

$$E_{el} = \frac{1}{2} T \ln \left(\frac{X_D}{X_M^2} \right), \tag{3.55}$$

where X_D, X_M are equilibrium concentrations of dimers and monomers, respectively. For the enzyme–DNA chimera of figure 3.17, the measured

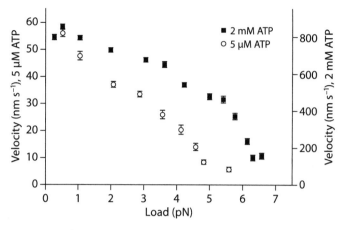

FIGURE 3.16. Velocity vs. applied force for the motor kinesin, at high and low ATP concentration (notice the different velocity scales for the two cases). Adapted from Visscher, Schnitzer, and Block (1999).

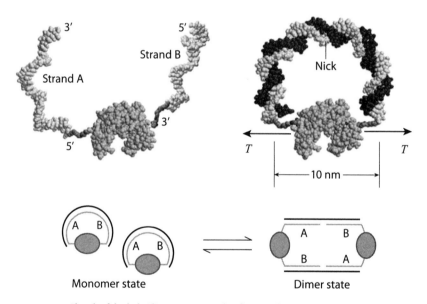

FIGURE 3.17. Sketch of the hybridization process that forms a chimera monomer with a nicked DNA spring. The enzyme depicted is GK. Two such monomers can form a dimer where the elastic energy is relaxed. Adapted from Zocchi (2009).

elastic energy is

$$E_{el} = 9.1 \pm 0.1 \, k_B T. \tag{3.56}$$

This is a large elastic energy, and it is at first surprising that it has so little effect on the enzymatic cycle (the effect of a 60 mer, nicked DNA spring is

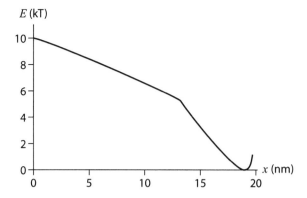

FIGURE 3.18. Calculated elastic energy vs. end-to-end distance x for a 60 mer DNA spring, nicked. The graph is a plot of eq. (3.51) using the parameter values $L = 19.8$ nm, $\tau_c = 27$ pN nm, $B = 200$ pN nm^2.

not visible for GK and quite small for RLuc; see figure 3.13). The reason is that, in this case, most of the elastic energy resides in the DNA spring, not in the enzyme. We can qualitatively see this by comparing the measured elastic energy of the chimera (3.56) with the elastic energy of the corresponding DNA spring, plotted in figure 3.18. If the enzyme was not deformed, the end-to-end distance of the DNA would be, from the geometry of this chimera, $s = 10$ nm. The elastic energy of just the DNA, for this value of end-to-end distance, is 6.6 kT. Thus by this estimate the elastic energy in the enzyme is only 2.5 kT.

We can view the enzyme–DNA chimera as a system of two coupled springs: a "DNA spring" that is compressed, and an "enzyme spring" that is elongated, with respect to their respective zero-force equilibria. In such a system, most of the elastic energy resides in the softer spring. For example, consider two linear springs (i.e., obeying Hooke's law) with different relaxed lengths X_1 and X_2, and different spring constants K_1 and K_2. Now couple the two springs so that they are constrained to have the same length (thus one spring will be elongated and the other compressed). If x is this common length, the elastic energy of the system is

$$E(x) = \frac{1}{2}K_1(x - X_1)^2 + \frac{1}{2}K_2(x - X_2)^2. \tag{3.57}$$

In mechanical equilibrium this energy is a minimum, i.e., $\partial E/\partial x = 0$, and we find that the energy of the system is

$$E = \frac{1}{2}(X_1 - X_2)^2 \frac{K_1 K_2}{K_1 + K_2}, \tag{3.58}$$

while the energies in the individual springs are in the ratio

$$\frac{E_1}{E_2} = \frac{K_2}{K_1}, \tag{3.59}$$

that is, there is more energy in the softer spring. In the next sections we examine in more detail the partition of elastic energy between the enzyme and the DNA spring.

3.5 Injection of Elastic Energy vs. Activity Modulation

Mechanical control of enzymes, and in general allosteric control, is a matter of putting the molecular structure under stress, that is, injecting elastic energy into the enzyme. To proceed with the discussion of the enzymatic cycle, we have to examine the relation of elastic energy injected to modulation of enzymatic activity. There is no simple answer to this question, but in order to extract some wisdom from the experimental results, it is useful to examine simple heuristic models.

Let us assume that different deformed states of the enzyme are associated with different "enzymatic activity" (we might think, for example, more specifically of the binding affinity for a substrate, or the catalytic rate k_{cat}). We examine the simplest, 1-D, continuum mechanical model where the enzyme has a zero temperature, unperturbed "length" ℓ, which can be stretched (but not compressed) at the cost of an elastic energy

$$E = \frac{1}{2}\kappa(x - \ell)^2, \quad x \geq \ell, \tag{3.60}$$

x being the end-to-end distance of this spring and κ the spring constant. For convenience, there are no states with $x < \ell$. We imagine that the "enzymatic activity" is maximum ($= 1$) for $x = \ell$ and decreases monotonically for $x > \ell$; to be specific, and for ease of computation, we assume that the ratio of the activity at elongation x to the maximum activity is

$$R = \exp\left(-\frac{x - \ell}{a}\right), \quad x \geq \ell. \tag{3.61}$$

The parameter a, which is a length, defines the elongation ($x = \ell + a$) where the activity drops significantly (here by a factor $1/e$). There is no justification for the specific form (3.61): we use it as an example, and seek to draw conclusions which, qualitatively, do not depend (much) on this particular choice. For notational ease, we introduce the zero-temperature parameters:

$$E_0 \equiv \frac{1}{2}\kappa a^2, \qquad E_{el} \equiv \frac{\sigma^2}{2\kappa}. \tag{3.62}$$

The term E_0 is the work done by the external field if the spring is elongated by a; E_{el} is the work done by the external mechanical force σ; both are

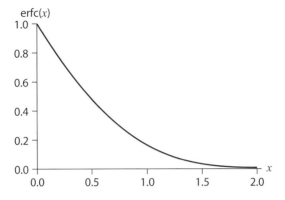

FIGURE 3.19. Graph of the complementary error function erfc defined by (3.67).

purely mechanical quantities. The statistical mechanics of this system is straightforward but instructive. The zero-force partition sum is

$$Z = \int_{\ell}^{+\infty} dx \, \exp\left(-\frac{\kappa(x-\ell)^2}{2T}\right) = \sqrt{\frac{2T}{\kappa}} \int_0^\infty dy \, \exp\left(-y^2\right) = \sqrt{\frac{\pi T}{2\kappa}}, \quad (3.63)$$

and the zero-force activity is

$$A = \frac{1}{Z} \int_{\ell}^{+\infty} dx \, \exp\left(-\frac{x-\ell}{a}\right) \exp\left(-\frac{\kappa}{2T}(x-\ell)^2\right). \quad (3.64)$$

With the variable substitution $\sqrt{\kappa/2T}\,(x-\ell) = y$, the integral transforms to

$$\sqrt{\frac{2T}{\kappa}} \int_0^{+\infty} dy \, \exp\left(-y^2 - \sqrt{\frac{2T}{\kappa a^2}}\, y\right). \quad (3.65)$$

We use the formula

$$\int_0^\infty dx \, \exp\left(-x^2 - \beta x\right) = \sqrt{\frac{\pi}{4}} \, \exp\left(\beta^2/4\right) \operatorname{erfc}\left(\frac{\beta}{2}\right), \quad (3.66)$$

where the complementary error function erfc is defined through the error function erf:

$$\operatorname{erf}(x) = \frac{2}{\sqrt{\pi}} \int_0^x e^{-u^2} \, du, \qquad \operatorname{erfc}(x) = 1 - \operatorname{erf}(x) = \frac{2}{\sqrt{\pi}} \int_x^{+\infty} e^{-u^2} \, du, \quad (3.67)$$

and obtain

$$A = \exp\left(\frac{T}{2\kappa a^2}\right) \operatorname{erfc}\left(\sqrt{\frac{T}{2\kappa a^2}}\right) = \exp\left(\frac{T}{4E_0}\right) \operatorname{erfc}\left(\sqrt{\frac{T}{4E_0}}\right), \quad (3.68)$$

where we have used definition (3.62). The graph of the complementary error function $\operatorname{erfc}(x)$ is plotted in figure 3.19. In the presence of an external force

σ, the new mechanical equilibrium length of the spring is $\ell + \sigma/\kappa$, and the energy is

$$E = \frac{1}{2}\kappa\left(x - \ell - \frac{\sigma}{\kappa}\right)^2 = \frac{1}{2}\kappa(x-\ell)^2 - \sigma(x-\ell) + \frac{\sigma^2}{2\kappa}, \qquad (3.69)$$

so the partition sum with an external force σ is

$$\begin{aligned} Z_\sigma &= \int_\ell^{+\infty} dx \, \exp\left(-\frac{\kappa}{2T}(x-\ell)^2 + \frac{\sigma}{T}(x-\ell)\right) \\ &= \sqrt{\frac{\pi}{4}}\sqrt{\frac{2T}{\kappa}}\,\exp\left(\frac{\sigma^2}{2\kappa T}\right)\mathrm{erfc}\left(-\sqrt{\frac{\sigma^2}{2\kappa T}}\right), \end{aligned} \qquad (3.70)$$

using the same manipulations as above. We have dropped the constant term $\sigma^2/2\kappa$ from the energy since it makes no difference to the ensemble averages that follow. However, we have to remember this "field energy" when we discuss energy or free energy changes under stress, as we will see later. The activity under stress is then

$$A_\sigma = \frac{1}{Z_\sigma}\int_\ell^{+\infty} dx \, \exp\left(-\frac{x-\ell}{a}\right)\exp\left(-\frac{\kappa}{2T}(x-\ell)^2\right)\exp\left(\frac{\sigma}{T}(x-\ell)\right). \qquad (3.71)$$

The integral can be transformed into

$$\sqrt{\frac{2T}{\kappa}}\int_0^\infty dy \, \exp\left(-y^2 + \left(\sqrt{\frac{2\sigma^2}{\kappa T}} - \sqrt{\frac{2T}{\kappa a^2}}\right)y\right), \qquad (3.72)$$

and using (3.66) once again, we obtain

$$\begin{aligned} A_\sigma = \frac{1}{Z_\sigma}\sqrt{\frac{\pi}{4}}\sqrt{\frac{2T}{\kappa}}\,&\exp\left(\frac{\sigma^2}{2\kappa T}\right)\exp\left(\frac{T}{2\kappa a^2}\right)\exp\left(-\frac{\sigma}{\kappa a}\right) \\ &\times \mathrm{erfc}\left[-\left(\sqrt{\frac{\sigma^2}{2\kappa T}} - \sqrt{\frac{T}{2\kappa a^2}}\right)\right], \end{aligned} \qquad (3.73)$$

or, using (3.70),

$$A_\sigma = \exp\left(\frac{T}{2\kappa a^2} - \frac{\sigma}{\kappa a}\right)\frac{\mathrm{erfc}\left[-\left(\sqrt{\sigma^2/2\kappa T} - \sqrt{T/2\kappa a^2}\right)\right]}{\mathrm{erfc}\left[-\sqrt{\sigma^2/2\kappa T}\right]}$$

$$= \exp\left(-\sqrt{\frac{E_{el}}{E_0}}\right)\exp\left(\frac{T}{4E_0}\right)\frac{\mathrm{erfc}\left[-\left(\sqrt{E_{el}/T} - \sqrt{T/4E_0}\right)\right]}{\mathrm{erfc}\left[-\sqrt{E_{el}/T}\right]}. \qquad (3.74)$$

The ratio of the activities with and without stress is, using (3.68),

$$\frac{A_\sigma}{A} = \exp\left(-\sqrt{\frac{E_{el}}{E_0}}\right)\frac{\mathrm{erfc}\left[\left(\sqrt{T/4E_0} - \sqrt{E_{el}/T}\right)\right]}{\mathrm{erfc}\left[-\sqrt{E_{el}/T}\right]\mathrm{erfc}\left[\sqrt{T/4E_0}\right]}. \qquad (3.75)$$

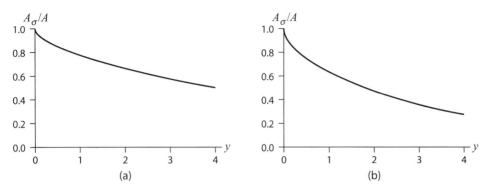

FIGURE 3.20. Plots of eq. (3.75) as a function of $y = E_{el}/T$, for a fixed value of $x = E_0/T$: (a) $x = 4$; (b) $x = 1$.

We examine (3.75) in the relevant region $0 < E_0, E_{el} <$ (a few kT). Qualitatively we expect the following: even at zero stress, the activity of the enzyme is reduced compared to the maximum possible value (which here is 1, corresponding to the spring at its mechanical equilibrium position), because thermal fluctuations allow the enzyme to explore states of lower activity. This effect is small if E_0 is relatively large, corresponding loosely speaking to a "stiff" enzyme (more precisely, an enzyme for which there is a relatively large energy difference between states of low activity and the state of maximum activity). In this case, it will take a relatively large injection of elastic energy, $E_{el} \sim E_0$, in order to modulate the activity significantly. In any case, whether E_0 is large or small (i.e., whether $E_0/T \gg 1$ or $E_0/T \ll 1$), a necessary condition to obtain significant modulation of activity is that we inject an elastic energy of at least E_0, that is, $E_{el} \sim E_0$ or larger. Figure 3.20 shows two plots of eq. (3.75) as a function of $y = E_{el}/T$, for two different fixed values of $x = E_0/T$. For $x = 4$ (i.e., $E_0 = 4kT$), we see that we need $y \approx 4$ ($E_{el} \approx 4\,kT$) in order to reduce the activity by a factor ~ 0.5. Figure 3.21(a) is a plot of (3.75) vs. $x = E_0/T$, for $y = x$ ($E_{el} = E_0$), showing that in any case we need $E_{el} > E_0$ for significant activity modulation. If we make the enzyme "soft" (i.e., $E_0 < T$: a small energy difference between states of low activity and the state of maximum activity), then we can obtain significant activity modulation while injecting a relatively small elastic energy. Figure 3.21(b) is a plot of (3.75) as a function of $x = E_0/T$, for $y = 4x$ ($E_{el} = 4E_0$). We see that, for example, if $E_0 = 0.2\,kT$, injecting a relatively small elastic energy $E_{el} = 4E_0 = 0.8\,kT$ still causes a decrease of activity by a factor ~ 0.5. However, there is a price to pay because such a "soft" enzyme is not optimized from the viewpoint of activity at zero stress. If $E_0 \ll T$, then even at zero stress the inactive states are accessed frequently, and the activity is reduced compared to a "stiff" enzyme. Figure 3.22 shows activity at zero stress, A, vs. $x = E_0/T$

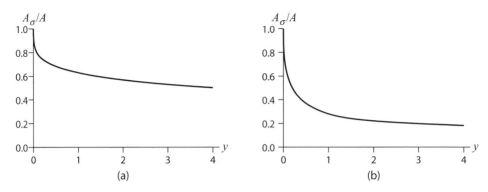

FIGURE 3.21. Plots of eq. (3.75) as a function of $x = E_0/T$: (a) $y = x$ ($E_{el} = E_0$); (b) $y = 4x$ ($E_{el} = 4E_0$).

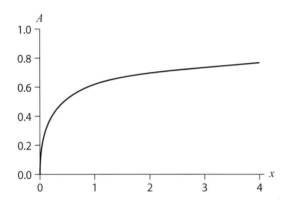

FIGURE 3.22. Graph of eq. (3.68) as a function of $x = E_0/T$.

(eq. (3.68)). We see that for the above example ($E_0 = 0.2T$), the zero-stress activity is a factor ~ 0.3 lower than the maximum value $A = 1$.

Thus from this heuristic model we extract the following ideas that may have some generality: enzyme mechano-chemical coupling may trade amplitude of allosteric modulation for overall enzymatic activity, in the sense that a "softer" enzyme, as defined above, can be modulated by a relatively small injection of elastic energy; however, this enzyme also has lower overall activity. Connected to this is the statement that, for a real enzyme, it seems reasonable that, in order to obtain significant modulation of activity through mechanical stress, the external force field has to provide an elastic energy $E_{el} \sim 1\,kT$ or larger. One might think that this is obvious since different states are visited with relative probabilities given by Boltzmann factors, but the preceding discussion shows some subtleties, and in fact the statement must be qualified, as we saw. To summarize this spring model, figure 3.23 shows

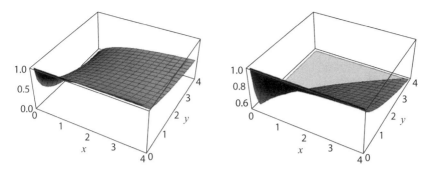

FIGURE 3.23. Left: graph of $(A_\sigma/A)(x, y)$ (eq. (3.75)) where $x \equiv E_0/T$, $y \equiv E_{el}/T$. Right: the same graph, cut by the plane $A_\sigma/A = 0.5$ to show the (x, y) region where there is substantial modulation of activity.

3-D plots of eq. (3.75). It is worth considering also the other thermodynamic functions of the model, because some features are at first surprising. We reinstate the constant term $\sigma^2/2\kappa$ which was dropped from the energy of the spring under stress, that is, we measure energies from the ground state, which is the state of mechanical equilibrium. Then the difference in the free energy of the spring with and without stress is

$$\Delta F = F_\sigma - F = -T \ln \frac{Z_\sigma}{Z} + \frac{\sigma^2}{2\kappa}, \tag{3.76}$$

where Z_σ, Z are given by (3.70) and (3.63). Thus we obtain

$$\frac{\Delta F}{T} = -\ln \left[\mathrm{erfc}\left(-\sqrt{\frac{E_{el}}{T}} \right) \right], \tag{3.77}$$

and similarly for the energy difference:

$$\Delta E = E_\sigma - E = T^2 \frac{\partial}{\partial T} \ln \left(\frac{Z_\sigma}{Z} \right) + \frac{\sigma^2}{2\kappa}. \tag{3.78}$$

From the definition of the complementary error function erfc it is easy to see that

$$\frac{d}{dx} \mathrm{erfc}(x) = -\frac{2}{\sqrt{\pi}} \exp\left(-x^2\right), \tag{3.79}$$

and carrying out the derivative in (3.78) we find

$$\frac{\Delta E}{T} = -\frac{1}{\sqrt{\pi}} \sqrt{\frac{E_{el}}{T}} \frac{\exp\left(-E_{el}/T\right)}{\mathrm{erfc}(-\sqrt{E_{el}/T})}, \tag{3.80}$$

while the entropy change is found from

$$\Delta S = \frac{\Delta E - \Delta F}{T}. \tag{3.81}$$

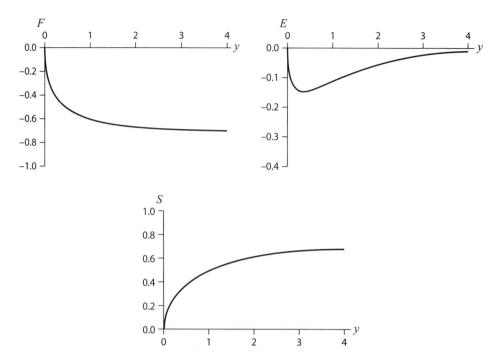

FIGURE 3.24. The free energy difference with and without stress, $\Delta F = F_\sigma - F$ (eq. (3.77)), the energy difference ΔE (eq. (3.80)), and the entropy difference ΔS (eq. (3.81)), plotted vs. $y \equiv E_{el}/T$ for the spring model (3.69).

These functions are plotted vs. $y \equiv E_{el}/T$ in figure 3.24. It seems counterintuitive that the energy should first decrease and then increase with the stress E_{el}, but the reason is that, with the restriction $x \geq \ell$, we have made our spring asymmetric. When a small stress is applied, some low energy "compressed" (with respect to the ground state) states become available that remove some probability from the higher energy "elongated" states, so the energy drops. As we increase the stress, new high energy "compressed" states become available and the energy rises. In contrast, the entropy is monotonically increasing because the stress always creates new "compressed" states; eventually for each elongated state there is a corresponding compressed state created, so the number of states doubles and $\Delta S \to \ln 2$ for $y = E_{el} \to \infty$ (figure 3.24). If, on the other hand, we remove the "field energy" $\sigma^2/2\kappa$ from the energy budget, then the entropy difference ΔS is unaffected because the external field carries no entropy, while the energy difference ΔE becomes monotonically decreasing:

$$\frac{\Delta E}{T} = -\frac{E_{el}}{T} - \frac{1}{\sqrt{\pi}}\sqrt{\frac{E_{el}}{T}}\frac{\exp\left(-E_{el}/T\right)}{\mathrm{erfc}(-\sqrt{E_{el}/T})}. \tag{3.82}$$

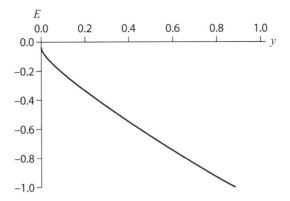

FIGURE 3.25. The energy difference with and without stress, ΔE, plotted vs. $y \equiv E_{el}/T$, not including the "field energy" (eq. (3.82)).

This choice corresponds to measuring energies always from the same state $x = \ell$ of the spring. The energy function (3.82) is plotted in figure 3.25. We used this spring model to introduce some ideas about the relation of mechanical stress to activity modulation. However, by itself the model does not describe even qualitatively the temperature dependence of the activity A. Equation (3.68) predicts a monotonically decreasing activity with temperature, whereas in reality, enzymatic activity first increases and then decreases for increasing temperature in the range where the enzyme is folded. The main contribution to the increase comes from an Arrhenius factor corresponding to the existence of a transition state for the chemical reaction, that is, the rate increases with temperature as

$$r = r_0 e^{-\Delta/T}, \tag{3.83}$$

where Δ is the barrier height to the transition state. The subsequent decrease is due to the proximity of the warm unfolding transition. Using (3.83), the ratio of activities at temperatures T_1 and T_2 is

$$\frac{r_1}{r_2} = \exp\left(-\Delta \frac{T_2 - T_1}{T_1 T_2}\right) \approx \exp\left(-\frac{\Delta}{T_0} \frac{T_2 - T_1}{T_0}\right), \tag{3.84}$$

where $T_0 = \sqrt{T_1 T_2}$. Switching for a moment to temperatures in degrees Kelvin, $T_0 \approx 300\,\mathrm{K}$ since T_1, T_2 are close to room temperature. A rate change by a factor 10 within a temperature range of $\sim 30\,\mathrm{K}$ (say from $T_1 = 300\,\mathrm{K}$ to $T_2 = 330\,\mathrm{K}$) is typical for enzymes, so we see from (3.84) that barrier heights Δ are relatively large, $\Delta \sim 10\,k_B T$ or more.

While the main temperature dependence of the enzymatic activity comes from barrier crossing, the (weaker) temperature dependence of the spring model can be changed with minor modifications. As an exercise, we may simply allow some "compressed" states, that is, replace the energy (3.60) with

$$E = \frac{1}{2}\kappa(x - \ell)^2, \quad x \geq \ell - a, \tag{3.85}$$

extending the same relation (3.61):

$$R = \exp\left((x - \ell)/a\right), \quad x \geq \ell - a. \tag{3.86}$$

We use the same length parameter a in the constraint for x and in the exponential function for R for simplicity. Now the compressed states $\ell - a < x < \ell$ have relative activity $R > 1$, so we expect that, as a function of temperature, the activity A will first increase (as these states become populated) and then decrease, that is, non-monotonic behavior. One can redo the statistical mechanics of the model in the form (3.85), (3.86). For example, the zero-stress partition sum is

$$Z = \int_{\ell-a}^{+\infty} dx \, \exp\left(-\frac{\kappa}{2T}(x - \ell)^2\right). \tag{3.87}$$

Using the formula

$$\int_{0}^{+\infty} dx \, \exp\left(-(ax^2 + \beta x)\right) = \sqrt{\frac{\pi}{4}} \frac{1}{\sqrt{a}} \exp\left(\frac{\beta^2}{4a}\right) \mathrm{erfc}\left(\frac{\beta}{2\sqrt{a}}\right), \tag{3.88}$$

we find

$$Z = \sqrt{\frac{\pi}{4}} \sqrt{\frac{2T}{\kappa}} \, \mathrm{erfc}\left(-\sqrt{\frac{\kappa a^2}{2T}}\right). \tag{3.89}$$

The result for the zero-stress activity is

$$A = \exp\left(\frac{T}{4E_0}\right) \frac{\mathrm{erfc}\left[-\left(\sqrt{4E_0/T} - \sqrt{T/E_0}\right)\right]}{\mathrm{erfc}\left(-\sqrt{E_0/T}\right)}, \tag{3.90}$$

and it is plotted in figure 3.26 as a function of T.

Instead of reasoning in the continuum mechanics limit, as we did above, one can also reason with discrete states of the enzyme; then the effect of the mechanical stress is to shift the energy levels of the system. In general, there will be many states, and one may work with a density of states, as we will see in chapter 4 with the kinetic measurements on myoglobin. Here we consider the opposite limit to a continuum description, which is just two states. This is unrealistic, but it is simplest and may be augmented later by adding more states. Consider then two states of the enzyme: we assign enzymatic activity 1 to one state, and 0 to the other. Now we need a prescription for how the stress shifts the energy levels (this is the "mechanics" of the system). We suppose that the energy shift is proportional to the stress. This is different from a spring, where the energy is proportional to the square of the stress.

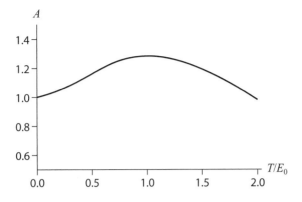

FIGURE 3.26. Graph of eq. (3.90), plotted vs. T/E_0.

It corresponds to a system that responds to deformations with a constant restoring force. This is unphysical for very small deformations but not so unphysical, for the enzyme, for larger deformations, as we will see later in this chapter and in chapter 4. With this prescription, we have the familiar problem of a spin $\frac{1}{2}$ system in a magnetic field. We assign activity $A = 1$ to the state $s_z = -\frac{1}{2}$, and activity $A = 0$ to the state $s_z = +\frac{1}{2}$; the magnetic field (in the z-direction) corresponds to the stress σ, that is, the two energy levels are $E = \pm m\sigma$ with m a constant; the upper sign is for the state $s_z = -\frac{1}{2}$. Then the activity A is related to the "magnetization" $M = 2m\langle s_z \rangle$ by

$$A = \frac{1 - M}{2}, \tag{3.91}$$

since in terms of the probabilities of the two states,

$$\frac{M}{m} = p\left(\frac{1}{2}\right) - p\left(-\frac{1}{2}\right) = 1 - 2p\left(-\frac{1}{2}\right) \quad \text{while } A = p\left(-\frac{1}{2}\right). \tag{3.92}$$

The partition sum of the system is

$$Z = \exp\left(\frac{m\sigma}{T}\right) + \exp\left(-\frac{m\sigma}{T}\right) \tag{3.93}$$

and the activity under stress is

$$A_\sigma = \frac{\exp(-m\sigma/T)}{\exp(m\sigma/T) + \exp(-m\sigma/T)} = \frac{1}{\exp(2m\sigma/T) + 1}. \tag{3.94}$$

Since the activity at zero stress is $A = \frac{1}{2}$, the ratio of activities with and without stress is

$$\frac{A_\sigma}{A} = \frac{2}{\exp(2m\sigma/T) + 1} = \frac{2}{\exp(2E_{el}/T) + 1}, \tag{3.95}$$

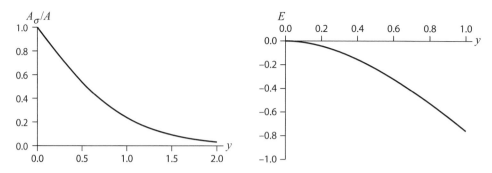

FIGURE 3.27. The activity ratio with and without stress, A_σ/A (eq. (3.95)) and the energy difference with and without stress, ΔE (eq. (3.96)), plotted vs. $y \equiv E_{el}/T$.

where we have introduced the "elastic energy" $E_{el} = m\sigma$. The other thermodynamic functions are also easily found; for example, the energy difference with and without stress is

$$\Delta E = E_\sigma - E = -m\sigma \tanh\left(\frac{m\sigma}{T}\right) = -E_{el} \tanh\left(\frac{E_{el}}{T}\right). \qquad (3.96)$$

The activity ratio (3.95) and the energy difference (3.96) are plotted in figure 3.27, and can be compared to the corresponding plots for the spring model (figures 3.20 and 3.25). We see that, qualitatively, these two opposite limits (continuum limit vs. two states) give the same result for the connection between elastic energy injected and activity modulation.

3.6 Connection to Nonlinear Dynamics: Two Coupled Nonlinear Springs

At the simplest level, an ASP-controlled enzyme (ASP = allosteric spring probe) may be viewed as a system of two coupled springs—at the molecular scale. We understand the DNA spring in some detail, so from measurements of the coupled system (figure 3.13) we should be able to say something about the mechanics of the other spring—the enzyme. The first statement is that in the enzyme–DNA spring construction (figure 3.4), the enzyme does not behave like a simple spring. We saw in the previous section that in order to obtain significant activity modulation mechanically (figures 3.10 and 3.12), the elastic energy injected into the enzyme must be of order $1\,kT$ or larger. Figure 3.28 shows calculated curves of the elastic energy in the enzyme, for the case of the RLuc chimera (figure 3.4) with the 60 mer and the 40 mer DNA springs. The model used for the DNA spring is eq. (3.51); the enzyme is supposed to obey Hooke's law with spring constant κ. The energy in the protein is plotted vs. κ, for different values of the critical bending torque

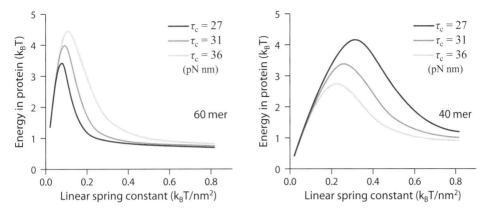

FIGURE 3.28. For the enzyme–DNA chimeras, calculated elastic energy injected in the enzyme assuming the latter behaves like a linear spring (i.e., follows Hooke's law). Adapted from Tseng and Zocchi (2014).

τ_c of the DNA spring. For the DNA spring with the nick, $\tau_c = 27\,\mathrm{pN\,nm}$ is the measured value; $\tau_c = 36\,\mathrm{pN\,nm}$ is meant to represent the case without nick: though this value is not measured directly, indications are that it lies somewhere in the range $31 < \tau_c < 40\,\mathrm{pN\,nm}$ (see chapter 2).

We see the following: For $\kappa \to \infty$, all curves approach asymptotically the value $E_p = \frac{1}{2}\,\mathrm{k_B T}$ (E_p = energy in the protein). This is just the thermal energy in the one degree of freedom that we are considering, corresponding to zero injected elastic energy. The reason is that if the enzyme is very stiff, all the elastic energy is in the DNA (see eq. (3.59)). To have at least $\sim 1\,\mathrm{k_B T}$ of injected elastic energy (i.e., $E_p > 1.5\,\mathrm{k_B T}$ in the plots), for the 60 mer spring, we need κ to be smaller than $\sim 0.4\,\mathrm{k_B T/nm^2}$ in all cases shown; for the 40 mer spring, we need κ to be smaller than $\sim 0.7\,\mathrm{k_B T/nm^2}$ in all cases. For the nicked 40 mer spring (figure 3.28, $\tau_c = 27\,\mathrm{pN\,nm}$), which has a substantial effect on the enzyme's activity (figure 3.12), $E_p > 1.5\,\mathrm{k_B T}$ corresponds to $\kappa < 0.4\,\mathrm{k_B T/nm^2}$. Next, in order to have a substantial modulation of activity between the nicked and non-nicked springs (figure 3.13), the injected elastic energy in the two cases should differ by $\sim 1\,\mathrm{k_B T}$ or more. Figure 3.28 (60 mer) says that this is possible only if $\kappa < 0.3\,\mathrm{k_B T/nm^2}$; figure 3.28 (40 mer) sets the upper limit at $\kappa < 0.6\,\mathrm{k_B T/nm^2}$.

Summarizing, if the enzyme obeys Hooke's law in the regime of forces applied by the DNA springs, then measurements of the activity under stress indicate that the enzyme's spring constant must be smaller than $\sim 0.6\,\mathrm{k_B T/nm^2}$ and probably of order $\kappa \sim 0.4\,\mathrm{k_B T/nm^2} \approx 1.6\,\mathrm{pN/nm}$. However, this is far too soft. For example, the corresponding rms thermal fluctuation in the diameter of the enzyme would be, by equipartition, $\langle x^2 \rangle^{1/2} = \sqrt{T/\kappa} \approx 1.5\,\mathrm{nm}$, whereas in reality these fluctuations are of order 1–2 Å.

Therefore, the spring constant of a hypothetical Hookean enzyme should be roughly a factor 100 larger: $\kappa \sim 40\,\mathrm{kT/nm^2}$, totally incompatible with the plots of figure 3.28.

Before proceeding, let us see how the curves of figure 3.28 are calculated. The energy function of the DNA spring, $E_{\mathrm{DNA}}(z)$, is given as a function of the end-to-end distance z by (3.51). For the protein, we assume a linear spring $E_{\mathrm{prot}}(x) = \frac{1}{2}\kappa x^2$, where x is the deformation. The elastic energy of the coupled system is

$$E(x) = E_{\mathrm{DNA}}(h + x) + E_{\mathrm{prot}}(x), \tag{3.97}$$

where x is the deformation of the enzyme and $h + x$ is the end-to-end distance of the DNA spring. Here, h is a geometric constant equal to the distance between the Cys residues where the DNA is attached on the surface of the enzyme plus the combined effective lengths of the cross-linkers used for the attachment. For the RLuc chimeras of figure 3.4, $h \approx 6.1\,\mathrm{nm}$. One computes the partition sum

$$Z = \int_0^{L-h} dx \, \exp\left(-\frac{E(x)}{T}\right) \tag{3.98}$$

(where L is the contour length of the DNA spring) using (3.97) and obtains the thermodynamic quantities in the usual way; in particular, the energy in the protein is

$$E_{\mathrm{p}} = \langle E_{\mathrm{prot}} \rangle = \frac{1}{Z} \int_0^{L-h} dx \, E_{\mathrm{prot}}(x) \exp\left(-\frac{E(x)}{T}\right). \tag{3.99}$$

This is the quantity plotted in figure 3.28.

Let us now see that, if we assume that the linear elasticity regime of the enzyme is limited by a softening transition, we can reconcile the mechanics with the measurements of activity under stress. We assume a nonlinearity in the enzyme's mechanics of the following form:

$$E_{\mathrm{prot}}(x) = \begin{cases} \dfrac{1}{2}\kappa x^2 & \text{for } 0 < x \le x_{\mathrm{e}}, \\[2mm] \dfrac{1}{2}\kappa x_{\mathrm{e}}^2 + f_{\mathrm{e}}(x - x_{\mathrm{e}}) & \text{for } x > x_{\mathrm{e}}. \end{cases} \tag{3.100}$$

The parameter f_{e} represents a constant (strain-independent) restoring force in the nonlinear regime. For convenience, we consider only elongation of the enzyme (i.e., $E_{\mathrm{prot}}(x) = +\infty$ for $x < 0$). The parameter x_{e} represents the critical strain where the spring softens (we take $\kappa x_{\mathrm{e}} > f_{\mathrm{e}}$). We expect x_{e} to be of order a few Å, f_{e} to be in the pN range, and the spring constant κ to be of order $100\,\mathrm{k_B T/nm^2}$, as we discussed in the previous section. The DNA spring is still described by eq. (3.51). Figure 3.29 shows the energy functions for this model, with the values $\kappa = 100\,\mathrm{k_B T/nm^2}$, $x_{\mathrm{e}} = 3\,\mathrm{Å}$.

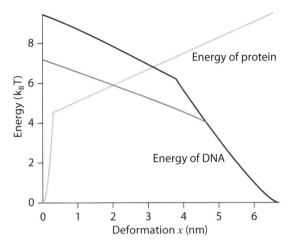

FIGURE 3.29. Energy functions for the enzyme and the DNA spring according to (3.100) and (3.51). The parameter values used for the enzyme spring are $\kappa = 100\,k_B T/nm^2$, $x_e = 3\,Å$, $f_e = 0.8\,k_B T/nm$. For the DNA, two different values of the critical bending torque τ_c are plotted (27 and 36 pN nm). Adapted from Tseng and Zocchi (2014).

One obtains the thermodynamic functions like before, using (3.100) in the partition sum (3.98). Figure 3.30 shows plots of the energy injected in the enzyme, $E_p = \langle E_{prot}(x)\rangle$, and the total energy of the enzyme–DNA system, $E_{tot} = \langle E_{prot}(x)\rangle + \langle E_{DNA}(x)\rangle$, as a function of the enzyme's restoring force in the nonlinear regime, f_e. The different curves correspond to different values of the DNA's critical bending torque τ_c, with $\tau_c = 27$ pN nm representing nicked DNA and $\tau_c = 36$ pN nm non-nicked DNA. Following the arguments of the previous discussion, we see from the plots for the 60 mer spring that the enzyme's restoring force f_e must be in the range $0.8 < f_e < 1.2\,k_B T/nm$, while the plots for the 40 mer spring give the approximate range $0.8 < f_e < 1.6\,k_B T/nm$. In conclusion, introducing a softening transition for the enzyme makes the mechanics compatible with the measured activity modulation under stress. The model accommodates a sufficiently large spring constant of the enzyme in the linear elasticity regime, $\kappa \sim 100\,k_B T/nm^2$. The enzyme's restoring force in the nonlinear regime must be of order $f_e \sim 1\,k_B T/nm = 4$ pN, which is of the same order as the forces produced by motor proteins. We will come back to this fundamental fact in chapter 4. For this value $f_e = 1\,k_B T/nm$, we see from figure 3.30 that the total elastic energy of the enzyme–DNA construct, in the case of nicked DNA ($\tau_c = 27$ pN nm), is approximately $E_{tot} = 9\,k_B T$, consistent with the measured values (see (3.56)). We will see in chapter 4 that this softening transition, here introduced heuristically, is directly observed dynamically in nano-rheology experiments.

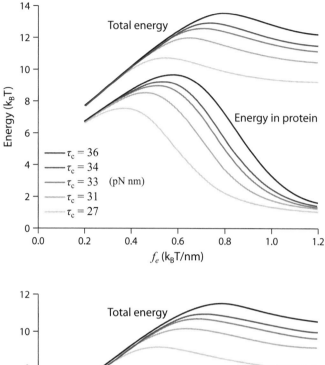

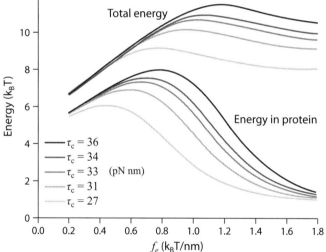

FIGURE 3.30. For the enzyme–DNA chimeras, calculated elastic energy injected in the enzyme assuming the softening transition (3.100) for the mechanics of the enzyme, with the parameter values $\kappa = 100\,\mathrm{kT/nm^2}$, $x_e = 3\,\text{Å}$. Left: 60 mer spring. Right: 40 mer spring. Adapted from Tseng and Zocchi (2014).

From the viewpoint of nonlinear physics, the collective behavior of coupled nonlinear springs is a fundamental topic. Indeed, the Fermi–Pasta–Ulam–Tsengou problem stands at the foundation of modern nonlinear dynamics. They considered different cases of nonlinearity, including a cubic

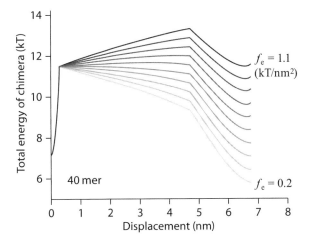

FIGURE 3.31. Total elastic energy for the enzyme–DNA construct according to the model (3.100), as a function of enzyme deformation. The different curves correspond to different values of the restoring force in the nonlinear regime, f_e (in steps of 0.1 k_B T/nm starting from the bottom curve $f_e = 0.2 k_B$ T/nm). The DNA energy function used is (3.51), and refers to a 40 mer spring.

term in the energy (quadratic in the force: this nonlinearity corresponds to a stiffening of the spring under elongation, and a softening under compression) and also a piecewise linear force, this latter case quite similar to (3.100). In our case, even the statics of the enzyme–DNA construction is nontrivial, for the following reason. Figure 3.31 shows the total elastic energy of the enzyme–DNA construct as a function of the enzyme deformation: this is simply adding the two energy functions of figure 3.29. The different curves are for different values of the enzyme's restoring force in the nonlinear regime, f_e. We see that relatively small changes in this parameter around the value $f_e \approx 0.7\,k_B$ T/nm result in a qualitative change of behavior of the system. For larger values of f_e, the system will spend most of the time near the undeformed state of the enzyme, and corresponding large deformation of the DNA. For smaller values of f_e, the opposite will occur. It is likely that in the real enzyme–DNA system, the proximity of a transition of this nature confers a particular sensitivity to small changes in the stiffness of the DNA spring, a characteristic that could be exploited for chemical detection purposes.

4

Dynamics of Enzyme Action

4.1 Introduction

The fundamental property of enzymes that concerns us in this chapter is their deformability. It allows enzymes to couple a chemical process to a cycle of deformations of the molecule, which can perform a task in the cell. This is the celebrated "molecular machine" aspect of enzymes. That enzymes must be deformable molecules was first understood by Koshland on purely biochemical grounds. Induced fit, allostery, and the propagation of mechanical stresses through the molecular structure are separate concepts only historically speaking. Conceptually, they are all manifestations of the fundamental property that enzymes are deformable molecules, and more specifically, that enzymes are generally deformed by binding–unbinding events.

The dynamics of enzyme deformability presents universal features when ensemble-averaged trajectories are examined. The mechanical response is viscoelastic. To explain viscoelasticity we consider a one-dimensional model which is also relevant to the experimental measurements on enzymes. We imagine a chunk of material compressed by a moving plate; we apply a force F (see figure 4.1). An elastic solid will respond with a deformation $x \propto F$:

$$x = F/\kappa. \tag{4.1}$$

A viscous liquid will respond with a deformation speed $\dot{x} \propto F$:

$$\dot{x} = F/\gamma, \tag{4.2}$$

where x is the position of the plate, κ is an elasticity parameter, and γ is a dissipation parameter. We suppose that masses and frequencies are small enough that inertial terms can be neglected. Then (4.1) and (4.2) are the equations of motion for the two cases (elastic/viscous). With an applied oscillatory force $F = F_0 e^{i\omega t}$, the solution for the elastic case is

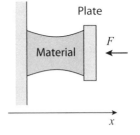

Plate

F

Material

x

$$x(t) = \frac{F_0}{\kappa} e^{i\omega t}, \qquad (4.3)$$

so the amplitude of the response,

$$|x| = \frac{F_0}{\kappa}, \qquad (4.4)$$

FIGURE 4.1. A rheology setup.

is constant (independent of frequency). This is the response of a spring. For the viscous case, the solution is

$$x(t) = \frac{F_0}{i\omega\gamma} e^{i\omega t}, \qquad (4.5)$$

so the amplitude decreases as $1/\omega$:

$$|x| = \frac{F_0}{\omega\gamma} \propto \frac{1}{\omega}. \qquad (4.6)$$

This is the response of a viscous flow. The dynamics (4.3) is non-dissipative: the work done by the force F over a cycle is

$$W = \int F\,dx = \int_0^{2\pi/\omega} F\dot{x}\,dt = 0 \qquad (4.7)$$

(we use the convention that the physical quantity is the real part of the corresponding complex quantity). In contrast, the dynamics (4.5) is dissipative:

$$W = \int_0^{2\pi/\omega} F\dot{x}\,dt = \int_0^{2\pi/\omega} F_0 \cos(\omega t) \left(\frac{F_0}{\gamma}\right)\cos(\omega t) = \frac{\pi F_0^2}{\gamma\omega} \neq 0, \quad (4.8)$$

a consequence of the response (4.5) being $\pi/2$ out of phase with the force ($\frac{1}{i} = e^{-i\pi/2}$).

A viscoelastic material responds to an applied force with a combination of (4.1) and (4.2):

$$\dot{x} = \frac{\dot{F}}{\kappa} + \frac{F}{\gamma}. \qquad (4.9)$$

For high frequencies $\omega \gg \omega_c = \kappa/\gamma$ this equation reduces to (4.1), whereas for low frequencies $\omega \ll \omega_c$ it reduces to (4.2). With an oscillatory applied force $F = F_0 e^{i\omega t}$, the solution of (4.9) is

$$x(t) = \frac{F_0}{i\omega\gamma}\left(1 + i\frac{\omega}{\omega_c}\right) e^{i\omega t}, \qquad (4.10)$$

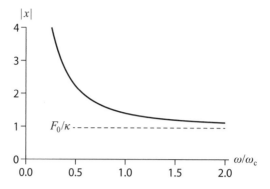

FIGURE 4.2. Amplitude of the viscoelastic response (Maxwell model). The graph is a plot of eq. (4.11).

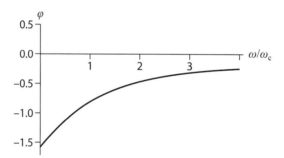

FIGURE 4.3. Phase of the viscoelastic response (in radians; Maxwell model, eq. (4.12)).

that is, an amplitude of deformation

$$|x| = \frac{F_0}{\gamma\omega}\sqrt{1 + \left(\frac{\omega}{\omega_c}\right)^2}. \tag{4.11}$$

The corner frequency $\omega_c = \kappa/\gamma$ separates the low frequency regime $\omega \ll \omega_c$, where the response (4.11) is that of a viscous flow ($|x| \approx F_0/(\gamma\omega)$), from the high frequency regime $\omega \gg \omega_c$, where the response is that of a spring ($|x| \approx F_0/\kappa$); see figure 4.2. The phase φ of the response (4.10) also transitions around ω_c: from (4.10) one finds

$$\varphi = -\arctan\left(\frac{\omega_c}{\omega}\right); \tag{4.12}$$

see figure 4.3. Thus the response is dissipative at low frequency and non-dissipative at high frequency.

In summary, a viscoelastic material behaves like an elastic solid at high frequency and like a viscous fluid at low frequency. Equation (4.9) is the Maxwell model of viscoelasticity.

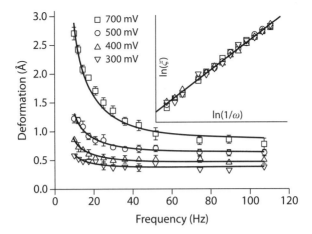

FIGURE 4.4. Mechanical response function measured for the enzyme guanylate kinase. The rms deformation amplitude is plotted vs. frequency of the applied force. Different data sets are for different amplitudes of the driving force. The lines are fits with the Maxwell model (4.11). Adapted from Qu, Landy, and Zocchi (2012).

4.2 Enzymes are Viscoelastic

The response (4.11) was discovered in the ensemble-averaged deformation mechanics of the enzyme guanylate kinase in 2011. In these experiments, the enzyme molecule takes the place of the chunk of material in figure 4.1, and the measurement is averaged over many replicas of the same system. An oscillatory force is applied to the plate of figure 4.1, and the amplitude (and phase) of the corresponding oscillatory deformation of the material (the enzyme molecule) is measured. The ensemble-averaged deformation of the enzymes is determined with sub-Å resolution; the response curves look like the example shown in figure 4.4. There are two important features: the plateau at "high" frequencies and the $1/\omega$ "divergence" at low frequencies. The corner frequency ω_c is of order $\omega_c \approx 2\pi \times 20\,\mathrm{Hz} \approx 100\,\mathrm{rad/s}$ in the experiments, and the measurements line up with the form (4.11) to within the experimental resolution of $\sim 0.2\,\text{Å}$. In short, for a fixed amplitude of the external force, the behavior vs. frequency is described by (4.9).

4.3 Nonlinearity of the Enzyme's Mechanics

The Maxwell model (4.9) represents a linear system, in which the amplitude of the response is proportional to the amplitude of the applied force, and the phase of the response is independent of the force. This model describes the

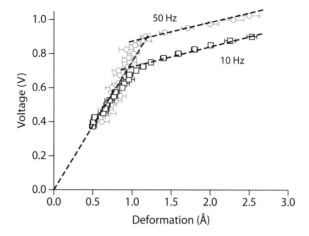

FIGURE 4.5. Mechanical response function measured for the enzyme guanylate kinase. Voltage is proportional to the amplitude of the applied oscillatory force, and is plotted vs. rms amplitude of the resulting deformation. The two data sets are for two different frequencies of the driving force (10 Hz and 50 Hz). Adapted from Wang and Zocchi (2011).

behavior of the enzyme vs. frequency but not vs. force. Enzyme mechanics is nonlinear, and displays a dynamic "softening transition" as a function of force F_0. This is seen in experiments where the frequency of the applied oscillatory force is kept constant while the amplitude of the force is varied; the response curve has a break at about 1 Å rms deformation, as shown in figure 4.5. The yield point moves with the frequency of the driving force, as seen in the figure, that is, the linear elasticity regime (where $|F| \propto |x|$) is extended at higher frequencies. In the $|x|-\omega$ plane, this nonlinearity appears as a shift in the corner frequency ω_c as F_0 is varied (figure 4.4). Namely, ω_c increases as F_0 increases.

At present there is no comprehensive model for this *viscoelastic transition*, for example in terms of a nonlinear equation that encompasses both the frequency response of (4.9) and the behavior vs. F shown in figure 4.5. What we do have is separate heuristic approaches for describing the two main experimental aspects of the transition. The frequency dependence is described by the viscoelastic model (4.9). The force dependence, specifically, the dependence of ω_c on the applied force (or, equivalently, the shift in the yield force with frequency; see figure 4.5) is described by the concept of barrier crossing in an energy landscape.

The model considers the conformational process in question (here, deforming the molecule beyond the yield strain) as similar to the problem of escape over a barrier for a particle in a thermal bath. The latter problem was discussed by Kramers in 1940, in the context of justifying the "transition state method" of expressing the rate of a chemical reaction. Specifically, the phenomena addressed by Kramers' theory are the dependence of these rates on temperature and "medium" viscosity. In our context, the Kramers model has been extended by Evans and Ritchie to explain the phenomenon that bond-rupture forces measured in AFM pulling experiments depend on the pulling speed.

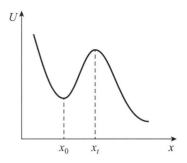

FIGURE 4.6. Energy landscape for the Kramers problem of escape over a potential barrier.

For the situation of figure 4.6, the rate of escape r for a particle initially in the bound state is

$$r = r_0 e^{-\Delta/T}, \tag{4.13}$$

where $\Delta = U(x_t) - U(x_0)$ is the barrier height. At first sight, this seems obvious: after all, in order to escape, the particle has to go through the position $x = x_t$ (in the context of chemical reactions $x = x_t$ is referred to as the "transition state"), and in a quasi-equilibrium situation, the probability of the particle finding itself in the neighborhood of the transition state is proportional to the corresponding Boltzmann factor, as in (4.13). However, this is not such a simple problem. The several assumptions implicit in the relation (4.13) restrict its range of validity. The relation (4.13) is meaningful insofar as it describes the main temperature dependence of r (i.e., with r_0 essentially temperature independent). One restriction must then be

$$\Delta/T > 1, \tag{4.14}$$

for in the opposite case the whole idea of a quasi-equilibrium process breaks down. The problem must then be stated differently (namely, what is the rate of escaping over the barrier for a particle initially placed at x_0?). This brings us to the form of r_0. According to the Kramers theory, and for the case of large viscosity (strong coupling to the medium),

$$r_0 = \frac{1}{2\pi} \frac{\sqrt{-U''(x_0)U''(x_t)}}{m} \tau, \tag{4.15}$$

where m is the mass of the particle and τ is the collision time representing the diffusion process for the particle (i.e., $\tau/m = \mu$ is the mobility of the particle, inversely proportional to the viscosity of the medium, $\mu = 1/(6\pi\eta R)$ in the Stokes regime; R is the size of the particle and η is the viscosity). We discussed this result in chapter 1. According to (4.15), r_0 depends on the shape of the potential, and $r_0 \propto 1/\eta$ (since $\tau \propto 1/\eta$). Coming back to (4.13), we can add the effect of a force f pulling the particle away from the bound state. The main effect is to lower the barrier by an amount $f\delta$ where $\delta = x_t - x_0$. So in the presence of a force field,

$$r_f = r_0 e^{-(\Delta - f\delta)/T} = r e^{f\delta/T}. \tag{4.16}$$

The escape rate increases exponentially with the force. Indeed, Evans and collaborators first pointed out and proved experimentally that because of (4.16), bond-rupture forces measured in AFM single molecule experiments depend on the pulling speed. In the context of the viscoelastic transition, we may consider the corner frequency ω_c as the rate of breaking the specific bond structure of the ground state conformation of the molecule (whereas small strains within the ground state conformation correspond to the linear elasticity response observed for $\omega \gg \omega_c$). In this case the same model (4.16) predicts

$$\omega_c = \omega_0 e^{f\delta/T}, \tag{4.17}$$

where f is the amplitude of the applied oscillatory force. Conversely, for the critical yield force f_c vs. frequency (figure 4.5) we expect

$$\ln \frac{\omega}{\omega_0} = \frac{f_c \delta}{T} \Rightarrow f_c = \frac{T}{\delta} \ln \frac{\omega}{\omega_0}. \tag{4.18}$$

The relations (4.17), (4.18) display the right qualitative trend for the force dependence of the corner frequency ω_c measured in the experiments, or equivalently, the frequency dependence of the yield force f_c. However, the (few) experimental measurements available indicate a stretched exponential dependence, that is, $\ln(\omega_c/\omega_0)$ increases faster than linearly in f.

4.4 Timescales

The viscoelastic transition is a materials property of enzymes, interesting because it is, presumably, general. Is viscoelasticity also relevant for the enzymatic cycle? For enzymes that display large conformational motion as part of the catalytic cycle, it seems reasonable that viscoelasticity controls the conformational dynamics. At the other end of the spectrum, enzymes that display only minimal conformational change during the catalytic cycle are basically similar to solid state catalysts, and though their rheological properties may also be viscoelastic, this materials property is likely irrelevant to their functioning. In the rest of the chapter we address the former group of enzymes only: let us call them, with Koshland, "induced-fit enzymes." Within this group, one must distinguish between the dynamics of conformational motion—set by the rheological properties of the molecule—and the overall rate of the enzymatic process. For some enzymes the chemical reaction step may be the slowest timescale (the rate limiting step), for others the conformational motion. Loosely speaking, for an optimized system these two timescales would be comparable. The viscoelastic transition provides a characteristic timescale $1/\omega_c$: conformational dynamics on timescales $\tau \geq 1/\omega_c$ is controlled by the viscoelastic materials properties of the molecule. Taking the example of guanylate kinase (GK), the corner frequency measured in the nano-rheology experiments is $\omega_c = 2\pi\nu_c \simeq 120\,\mathrm{rad/s}$. However, we saw that this rate depends on the applied force, in this case, the forces exerted by GMP on the enzyme structure, which drive the open-to-closed conformational transition of the enzyme. The rate for the chemical reaction step, k_{cat}, is in the range $100\text{--}300\,\mathrm{Hz}$. The value of k_{cat} is obtained by analyzing the dependence of the overall rate of the enzymatic reaction on substrate concentration (see chapter 3). One assumes Michaelis–Menten (MM) kinetics, and k_{cat} thus measured should perhaps more properly be taken as a lower bound for the rate of the chemical reaction step. Similarly, the measured k_{cat} represents a lower bound for the rate of the open-to-closed conformational transition. Putting everything together, from these timescales we evince that it is plausible that, for GK, the open-to-closed conformational motion during the catalytic cycle happens across the viscoelastic transition seen in the nano-rheology experiments. Plausible, but not proven beyond doubt. On the other hand, the yield deformation in the nano-rheology experiments is $1\,\text{Å}$ rms amplitude or $2.8\,\text{Å}$ peak to peak. For this particular molecule (GK from TB), the corresponding deformation obtained by comparing the X-ray structures of the open and closed forms is approximately $5\,\text{Å}$ (see figure 4.7). So again it appears that the open-to-closed conformational transition during the enzymatic cycle happens in the viscoelastic regime.

Finally, whether the mechanical steps or the chemical step are rate limiting in the case of GK is not quite clear: the timescales could be comparable, or the mechanics could be faster.

4.5 Enzymatic Cycle and Viscoelasticity: Motors

We can now give a representation of the enzymatic cycle based on the viscoelastic transition. Our aim is to give a description that is general enough and idealized to an extent that it clarifies the important concepts. Real, individual enzymes will depart from this ideal cycle in various ways. Ideally, our description would be in relation to real enzymatic cycles, similar to the relation of the Carnot cycle to real engines.

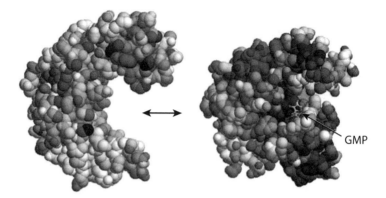

FIGURE 4.7. Two structures of the enzyme guanylate kinase (GK) from yeast. On the left is the structure in the absence of ligands; on the right, the structure with GMP bound. The enzyme is about 4 nm across. This is an enzyme of the induced-fit type: GMP binding drives a large conformational change of the molecule. The structures are PDB codes 1EX6, 1EX7. Adapted from Qu, Landy, and Zocchi (2012).

To fix ideas, we have in mind the example of GK and we consider only one substrate for simplicity. The process we consider is then as follows:

1. The substrate "binds" (to the "open" form of the enzyme); this step is diffusion limited and gives rise to the dependence of the enzymatic speed on substrate concentration, captured by the Michaelis–Menten description: speed $\propto 1/(K/[S] + 1)$, where $[S]$ is the substrate concentration and K is the Michaelis–Menten constant.
2. The forces exerted by the substrate on the different parts of the enzyme (electrostatic, hydrogen-bonding, entropic forces arising from a different organization of the hydration layer) drive the open-to-closed conformational motion.
3. The chemical reaction takes place.
4. The products unbind.
5. The internal restoring force of the structure, in the absence of ligands, drives the closed-to-open conformational motion; the enzyme comes back to the initial state (or close to it).

It is good to remember that this chronology of events is, itself, a model. The real physical system moves in a phase space of dimension $\sim 10^4$ (consisting of the momenta and coordinates of all the atoms of the enzyme and substrate plus one or two shells of water molecules). The chronology above is a qualitative representation of how the *ensemble-averaged* trajectory of the system might look during the enzymatic cycle, as far as the position of the atoms is concerned. Other ensemble-averaged quantities of interest during the cycle are $\langle \vec{p}^{\,2} \rangle$, $\langle \vec{x}^{\,2} \rangle$ for each atom ($\vec{p}$ is the momentum, $\vec{x}$ is the position).

The cycle is very slow compared to all thermalization processes within the enzyme (the collision time for atoms is a fraction of a ps; the heat diffusion time across the enzyme is $\tau \sim \frac{\delta^2}{D} \simeq \frac{(4 \times 10^{-7}\,\text{cm})^2}{10^{-3}\,\text{cm}^2/\text{s}} \simeq 10^{-11}\,\text{s}$). Thus it is reasonable to assume that $\langle \vec{p}^2 \rangle = 3T/m$ for each atom, that is, the structure is in thermal equilibrium throughout the cycle. On the other hand, $\langle \vec{x}^2 \rangle$ for each

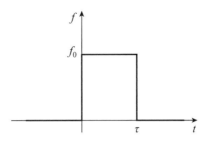

FIGURE 4.8. The force eq. (4.21).

atom will be varying in general throughout the cycle, reflecting the changing local interaction strengths ("spring constants") as the molecule deforms. We are saying, simply, different parts of the molecule may be more or less "floppy" at different stages of the enzymatic cycle. Correspondingly, there are entropy terms in the free energy of the system that vary in the course of the cycle; these terms are not small (see the argument about the entropy of a spring, below).

Coming back to the chronology $(1) \rightarrow (5)$ above, note also that steps (4) and (5) may actually occur in reverse order, or even "at the same time."

Now we cast the chronology $(1) \rightarrow (5)$ in terms of viscoelastic mechanics. Step (1) defines the initial time $(t = 0)$ for the cycle; $x(t)$ describes the deformation of the enzyme, and obeys viscoelastic dynamics:

$$\dot{x} = \frac{\dot{f}}{\kappa} + \frac{f}{\gamma}, \tag{4.19}$$

where f is the force,

$$f = -\frac{\partial F}{\partial x}, \tag{4.20}$$

and $F = F(x)$ is the free energy of the system. The energy F also depends (in a stepwise fashion) on the presence or absence of bound substrates and products. We consider step (2), and take for the force f the simplest form, namely a step function:

$$f = \begin{cases} f_0 & \text{for } 0 < t < \tau, \\ 0 & \text{otherwise} \end{cases} \tag{4.21}$$

(see figure 4.8). Namely, the substrate binds at time $t = 0$; it "pulls" the structure towards the closed conformation with a constant force; at time $t = \tau$ the closed conformation is reached, which is a free energy minimum (zero force). From (4.21),

$$\dot{f} = f_0[\delta(t) - \delta(t - \tau)], \tag{4.22}$$

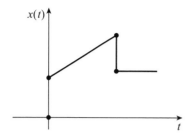

FIGURE 4.9. The response eq. (4.26).

and thus,

$$\dot{x} = \frac{f_0}{\kappa}[\delta(t) - \delta(t - \tau)] + \frac{f_0}{\gamma} \quad (0 \leq t \leq \tau).$$
(4.23)

Integrating (4.23) from 0 to t, we find, for $0 < t < \tau$,

$$x(t) - x(0) = \frac{f_0}{\kappa} + \frac{f_0}{\gamma}t, \quad (4.24)$$

while integrating (4.23) from $\tau - \varepsilon$ to $\tau + \varepsilon$ gives

$$x(\tau + \varepsilon) - x(\tau - \varepsilon) = -\frac{f_0}{\kappa}. \quad (4.25)$$

Finally, the trajectory is

$$x(t) = \begin{cases} 0 & \text{for } t < 0, \\ \frac{f_0}{\kappa} + \frac{f_0}{\gamma}t & \text{for } 0 < t < \tau, \\ \frac{f_0}{\gamma}\tau & \text{for } t > \tau \end{cases} \quad (4.26)$$

(see figure 4.9). Thus the viscoelastic response to an impulsive force, such as we envision the stress on the enzyme structure resulting from substrate binding to be, is an elastic deformation f_0/κ followed by a "flow" of amplitude $(f_0/\gamma)\tau$, which is larger than the elastic deformation if $(f_0/\gamma)\tau > f_0/\kappa \Rightarrow \tau > \gamma/\kappa$, or $\tau > 1/\omega_c$ in the language of the nano-rheology experiments. Thus the dynamics of the conformational cycle is dominated by the viscoelastic transition (as opposed to the linear elasticity regime) if the rate of the open-to-closed transition is

$$\text{rate} = \frac{1}{\tau} < \omega_c. \quad (4.27)$$

To describe the entire cycle, we introduce a restoring force of the enzyme, g, which in the absence of substrates or products drives the enzyme towards the ligand-free (open) conformation. For simplicity, g is also constant:

$$g = \begin{cases} -g_0 & \text{for } x > 0, \\ 0 & \text{for } x = 0, \\ g_0 & \text{for } x < 0. \end{cases} \quad (4.28)$$

Now when the substrate binds, the total force is $f_0 - g_0$. The complete cycle is represented as

$$f(t) = \begin{cases} f_0 - g_0 & \text{for } 0 \leq t < \tau_1, \\ -g_0 & \text{for } \tau_1 \leq t < \tau_1 + \tau_2, \\ 0 & \text{otherwise.} \end{cases} \quad (4.29)$$

Equation (4.29) is the force profile to be used in (4.19); τ_1 is the duration of step (2) (open-to-closed transition), and τ_2 is the duration of step (5) (relaxation of the enzyme structure back to the open conformation, in the absence of ligands). Integrating (4.19) with the force profile (4.29), noting that

$$\dot{f} = (f_0 - g_0)\delta(t) - f_0\delta(t - \tau_1) + g_0\delta(t - \tau_1 - \tau_2), \quad (4.30)$$

one obtains the trajectory

$$x(t) = \begin{cases} \frac{f_0 - g_0}{\kappa} + \frac{f_0 - g_0}{\gamma}t & \text{for } 0 \leq t < \tau_1, \\ -\frac{g_0}{\kappa} + \frac{f_0}{\gamma}\tau_1 - \frac{f_1}{\gamma}t & \text{for } \tau_1 \leq t < \tau_1 + \tau_2, \\ 0 & \text{otherwise.} \end{cases} \quad (4.31)$$

We can write the timescales τ_1, τ_2 in terms of the forces f_0, g_0. The amplitude of the conformational motion is

$$x_{\max} = x(\tau_1^-) = \frac{f_0 - g_0}{\kappa}(1 + \omega_c\tau_1) \quad (4.32)$$

or

$$\tau_1 = \frac{1}{\omega_c}\left(\frac{\kappa x_{\max}}{f_0 - g_0} - 1\right), \quad (4.33)$$

while in order that the enzyme be back to the original state at the end of the cycle (i.e., $x(t > \tau_1 + \tau_2) = 0$) we must have

$$\tau_2 = \frac{f_0 - g_0}{g_0}\tau_1. \quad (4.34)$$

In (4.33), $\kappa x_{\max}$ is the elastic force if the conformational motion were all in the elastic regime, while $f_0 - g_0$ is the actual force on the enzyme. Due to the viscoelastic transition (see figure 4.5), $\kappa x_{\max}/(f_0 - g_0) \gg 1$. Then from (4.33), (4.34) we obtain for the rates of the forward and backward conformational motion,

$$\begin{cases} \frac{1}{\tau_1} = \omega_c\frac{f_0 - g_0}{\kappa x_{\max}}, \\ \frac{1}{\tau_2} = \omega_c\frac{g_0}{\kappa x_{\max}}. \end{cases} \quad (4.35)$$

The rates for the mechanical motion are set by the corner frequency of the viscoelastic transition.

This mechanical cycle can be seen as an engine. To explore its dynamics we apply a load f_L to the backward (closed-to-open) motion, so the force (4.29) is now

$$f(t) = \begin{cases} f_0 - g_0 & \text{for } 0 \le t < \tau_1, \\ -g_0 + f_L & \text{for } \tau_1 \le t < \tau_1 + \tau_2, \\ 0 & \text{for } t \ge \tau_1 + \tau_2, \end{cases} \tag{4.36}$$

which gives the trajectory

$$x(t) = \begin{cases} \frac{f_0 - g_0}{\kappa} + \frac{f_0 - g_0}{\gamma} t & \text{for } 0 \le t < \tau_1, \\ -\frac{g_0}{\kappa} + \frac{f_L}{\kappa} + \frac{f_0 - f_L}{\gamma} \tau_1 - \frac{g_0 - f_L}{\gamma} t & \text{for } \tau_1 \le t < \tau_1 + \tau_2. \end{cases} \tag{4.37}$$

Now the condition for τ_2 is (using $\omega_c \tau_1 \gg 1$, as before)

$$\tau_2 = \frac{f_0 - g_0}{g_0 - f_L} \tau_1, \tag{4.38}$$

and finally (4.35) is replaced by

$$\begin{cases} \frac{1}{\tau_1} = \omega_c \frac{f_0 - g_0}{\kappa x_{max}}, \\ \frac{1}{\tau_2} = \omega_c \frac{g_0 - f_L}{\kappa x_{max}}. \end{cases} \tag{4.39}$$

From (4.39) we identify g_0 as the "stall force" of the mechanical cycle (for $f_L \to g_0$, $1/\tau_2 \to 0$):

$$g_0 = f_{stall}. \tag{4.40}$$

This quantity is directly measured in single molecule experiments on processive enzymes, notably motor proteins, and could in principle be measured more generally for any enzyme of the "induced-fit" kind.

In the case $\tau_2 \gg \tau_1$, which may be typical for motors under load, the total duration of the mechanical cycle is

$$\tau = \tau_1 + \tau_2 \approx \tau_2 = \frac{f_0 - g_0}{g_0 - f_L} \tau_1, \tag{4.41}$$

or equivalently an overall rate of

$$R = \frac{1}{\tau_2} = \omega_c \frac{g_0 - f_L}{\kappa x_{max}}. \tag{4.42}$$

Thus in this model the overall rate of the cycle goes to zero linearly with the load f_L. The maximum rate R_{max} is obtained for $f_L = 0$, and from (4.42), remembering that $\omega_c = \kappa/\gamma$, one finds

$$R_{max} = \frac{f_{stall}}{\gamma x_{max}}, \tag{4.43}$$

that is, a general relation between the zero load rate of the cycle, the amplitude of the conformational motion, and the dissipation parameter γ.

The work done on the enzyme by the force arising from the binding of the substrate is $W_1 = f_0 x_{max}$, while the work delivered by the enzyme against the load is $W_3 = f_L x_{max}$, giving an efficiency

$$\eta = \frac{W_3}{W_1} = \frac{f_L}{f_0}. \tag{4.44}$$

In this scenario $f_L < g_0 < f_0$, and the efficiency can in principle be up to 1, but at the price of an infinitely slow cycle (recall that for $f_L = g_0$ the backward, closed-to-open motion stalls, while for $g_0 = f_0$ the forward motion stalls). Also note that there is no "hydrodynamic" dissipation if one moves infinitely slowly.

Let us come back to the duration of the cycle eq. (4.41) without assumptions on the relative magnitudes of τ_1 and τ_2. Formula (4.41) then reads

$$\tau = \tau_1 + \tau_2 = \frac{1}{\omega_c} \frac{f_0 - f_L}{g_0 - f_L} \frac{\kappa x_{max}}{f_0 - g_0}, \tag{4.45}$$

or, in terms of the efficiency $\eta = f_L/g_0$,

$$\tau \omega_c = \frac{\kappa x_{max}}{f_0} \frac{1 - \eta}{(1 - g_0/f_0)(g_0/f_0 - \eta)}, \tag{4.46}$$

where

$$f_L < g_0 < f_0 \Rightarrow \eta < \frac{g_0}{f_0} < 1. \tag{4.47}$$

High efficiency ($\eta \approx 1$) is obtained for $f_L \approx g_0 \approx f_0$, in which case the cycle is slow. Since $g_0/f_0 - \eta < 1 - \eta$ in general,

$$\tau \omega_c > \frac{\kappa x_{max}}{f_0} \frac{1}{1 - g_0/f_0}. \tag{4.48}$$

This equation shows that the duration of the cycle, in units of $1/\omega_c$, is set by $\kappa x_{max}/f_0 \gg 1$ times a factor $(1 - f_{stall}/f_0)^{-1}$ (recall that $g_0 = f_{stall}$) which is of order 1 for "small" stall forces but can be large if the stall force is large. Let us find the condition that maximizes the rate of the cycle (minimizes τ) for a given load: setting $\partial \tau/\partial g_0 = 0$ in (4.45), considering f_L, f_0 fixed, we find

$$g_0 = \frac{f_0 + f_L}{2}, \tag{4.49}$$

so an enzyme set up for maximum speed under zero load has $g_0 = f_0/2$ and thus an efficiency

$$\eta = \frac{f_L}{f_0} < \frac{g_0}{f_0} = \frac{1}{2}, \tag{4.50}$$

or

$$\eta_{max} = \frac{1}{2}. \tag{4.51}$$

The zero load rate is then

$$\tau\omega_c = \frac{f_0}{f_0/2}\frac{\kappa x_{max}}{f_0/2} = 4\frac{\kappa x_{max}}{f_0}, \tag{4.52}$$

while with a load (using (4.49) in (4.45)),

$$\tau\omega_c = 4\frac{\kappa x_{max}}{f_0 - f_L} = 2\frac{\kappa x_{max}}{f_{stall} - f_L}. \tag{4.53}$$

How do the conclusions from this relaxation model compare to experiments? We take the case of kinesin, a motor protein that "walks" along microtubules and is able to pull against a load; its mechano-chemical properties have been studied extensively in beautiful single molecule experiments by the Block group and others. In terms of the simplified process of this section, step (1) corresponds to ATP binding, step (2) is driven by ATP binding, or binding to the microtubule, or both, step (3) is ATP hydrolysis, and steps (4) and (5) correspond to the "power stroke." The kinesin head detaches from the microtubule after step (5).

Kinesin is certainly a more complex enzyme compared to, say, guanylate kinase: it walks on microtubules! However, some of the fundamental aspects of the dynamics need not be very different. Both bind ATP and a second ligand—GMP for guanylate kinase, the microtubule for kinesin. The binding–unbinding of ligands drives a cycle of large conformational changes. The difference in the chemical potential of substrates and products drives the cycle in one direction. Obviously, kinesin has some "extra features," including "lever arms" that amplify the conformational motion, and two interacting heads that ensure the molecule as a whole remains bound to the microtubule while stepping. At our level of description, we regard these complications as details that may or may not obscure the fundamental dynamics, which rests on the materials properties of the protein.

In the language of eq. (4.42), the speed of the motor (at saturating ATP concentrations) is

$$v = Rx_{max} = \omega_c\frac{f_{stall} - f_L}{\kappa} = \frac{1}{\gamma}(f_{stall} - f_L) \tag{4.54}$$

(we use (4.42) rather than (4.53) in the belief that, for kinesin, $\tau_2 \gg \tau_1$, i.e., the cycle is fast–slow). The speed goes to zero linearly with the load force f_L.

However, the experimental force–velocity curves for kinesin and other motors are not quite linear.

From (4.54) we have the relation for the maximum speed of the motor ($f_L = 0$),

$$v_{max} = \frac{f_{stall}}{\gamma} \qquad (4.55)$$

(this result is independent of "lever arm length"). For kinesin, $f_{stall} \approx 10\,\mathrm{pN}$ and $v_{max} \approx 1\,\mu\mathrm{m/s}$; according to (4.55) this gives $\gamma = 10^{-2}\,\mathrm{g/s}$.

4.6 Internal Dissipation

The value of the dissipation parameter γ can be extracted—at least as an order of magnitude—from the mechanical "nano-rheology" measurements represented in figure 4.4. The parameter measured in the experiments is the corner frequency $\omega_c \approx 100\,\mathrm{rad/s}$; with the interpretation $\omega_c = \kappa/\gamma$, this is a measurement of γ given a value for the elastic parameter κ. A consensus value for κ is hard to pinpoint from the relatively few experimental measurements, but as an order of magnitude this elastic constant lies in the range $10\text{–}10^3\,\mathrm{pN/nm}$. Choosing $\kappa = 100\,\mathrm{pN/nm}$ one obtains

$$\gamma = \frac{\kappa}{\omega_c} = 1\,\mathrm{g/s}. \qquad (4.56)$$

There are two further, independent mechanical measurements of the dissipation parameter γ: the original AFM indentation experiments by the Hansma group, reporting a value $\gamma \approx 10^{-2}\,\mathrm{g/s}$, and an experiment by this author's group using a voltage-gated ion channel as both the rheometer and the material under study; this study reports a value $\gamma \approx 1\,\mathrm{g/s}$.

For a macroscopic, isotropic material, one represents elasticity and dissipation by Young's modulus Y and viscosity η. The enzyme is a heterogeneous nanoparticle and, as a material, it cannot be scaled up; therefore the concepts of Y and η are not operationally well defined. However, one can still talk of Y and η as an order of magnitude. To relate κ and γ to Y and η one needs a length scale. For an order of magnitude calculation, the appropriate length scale is the square root of the contact area in the mechanical experiment used to measure κ and η. The keyword here is the *Hertz* model for the contact stresses. Using, then, the length scale

$$a = \sqrt{R\delta}, \qquad (4.57)$$

where R is the radius of the enzyme and δ is the indentation in the experiments, we have

$$Y \sim \frac{\kappa}{a}, \qquad \eta \sim \frac{\gamma}{a} \qquad (4.58)$$

for Young's modulus and viscosity, respectively. Since $R \approx 2\,\text{nm}$, $\delta \approx 1\,\text{Å}$, we find $a \approx 0.5\,\text{nm}$; let us take $a = 1\,\text{nm}$ for simplicity in our estimates. Using $\kappa = 100\,\text{pN/nm}$ we find

$$Y = \frac{\kappa}{a} = \frac{100\,\text{pN/nm}}{1\,\text{nm}} = 10^9\,\frac{\text{dyn}}{\text{cm}^2} = 10^8\,\frac{\text{N}}{\text{m}^2}. \tag{4.59}$$

The mks unit of pressure is the Pascal (Pa), so the above estimate is $Y \sim 100\,\text{MPa}$. The range $\kappa \sim 10\text{--}10^3\,\text{pN/nm}$ thus corresponds to the range $Y \sim 10\,\text{MPa--}1\,\text{GPa}$ for Young's modulus. Similarly, using $\gamma = 0.1\,\text{g/s}$ we find

$$\eta = \frac{\gamma}{a} = \frac{0.1\,\text{g/s}}{10^{-7}\,\text{cm}} = 10^6\,\frac{\text{g}}{\text{cm s}}. \tag{4.60}$$

In cgs units, the viscosity of water is $\eta_\text{w} \sim 10^{-2}\,\text{g/(cm s)}$, so as an order of magnitude, the internal viscosity of the enzyme is

$$\eta \sim 10^8 \eta_\text{w}. \tag{4.61}$$

The enzyme is, after all, a solid! The range $\gamma \sim 10^{-2}\text{--}1\,\text{g/s}$ thus corresponds to the range $\eta \sim 10^5\text{--}10^7\,\text{g/(cm s)} = 10^7\text{--}10^9 \times \eta_\text{w}$. To summarize, the operationally well-defined dissipation parameter γ, associated with enzyme deformations, lies in the range

$$\gamma \sim 10^{-2}\text{--}1\,\text{g/s}. \tag{4.62}$$

4.7 Origin of the Restoring Force g

The restoring force is $g = -\partial F / \partial x$, where $F(x)$ is the free energy in the absence of ligands. The free energy $F(x)$, and therefore g, has an energy and an entropy component. The energy part comes from rearranging soft bonds (e.g., hydrogen bonds) as x (the conformation) changes. The entropy part comes from the following effect.

If the interactions that "hold" an atom or group of atoms (e.g., an amino-acid side chain) "weaken," the entropy increases, while if the interactions "strengthen," the entropy of that atom or group decreases. To see this, consider the entropy of an atom held in position by a spring, at temperature T. The Hamiltonian is

$$H = \frac{p^2}{2m} + \frac{1}{2}Kx^2, \tag{4.63}$$

where m is the mass of the atom and K is the spring constant. Using a classical calculation in the high temperature regime $T > \hbar\omega$, $\omega = \sqrt{K/m}$ (and in 1-D for simplicity), the partition sum is

$$Z = \int_{-\infty}^{+\infty} \frac{dp\,dx}{h} \exp\left(-\left(\frac{p^2}{2m} + \frac{1}{2}kx^2 \right)/T \right), \tag{4.64}$$

which is easily calculated to be

$$Z = \frac{T}{\hbar\omega}, \quad \text{where } \omega^2 = \frac{k}{m}. \tag{4.65}$$

The free energy, energy, and entropy are therefore

$$F = -T \ln Z = -T \ln\left(\frac{T}{\hbar\omega}\right), \tag{4.66}$$

$$E = T^2 \frac{\partial \ln Z}{\partial T} = T, \tag{4.67}$$

$$S = \frac{E-F}{T} = \ln\left(\frac{T}{\hbar\omega}\right) + 1, \tag{4.68}$$

or

$$S = \ln\left(\frac{T}{\hbar}\sqrt{\frac{m}{k}}\right) + 1. \tag{4.69}$$

We see that a stiffer spring (larger k) corresponds to a smaller entropy. This is *not* a small effect: the log term in (4.69) is typically larger than 1, as we can see from the following estimate.

An effective value for k can be found from the thermal equilibrium condition $\frac{1}{2}k\langle x^2\rangle = \frac{1}{2}T \Rightarrow k = \frac{T}{\langle x^2\rangle}$, knowing (from Debye–Waller factors for example) that typical fluctuations of atoms or groups of atoms (e.g., amino-acid side chains) in the folded structure at room temperature are $\langle x^2\rangle \sim (0.5\,\text{Å})^2$. Let us then take, numerically, $\langle x^2\rangle = r_0^2$ (where $r_0 = \hbar^2/(m_e e^2)$, the Bohr radius). The argument of the log in (4.69) is then

$$\frac{T}{\hbar}\sqrt{\frac{m}{k}} \simeq \frac{T}{\hbar}\sqrt{\frac{mr_0^2}{T}} = \sqrt{\frac{Tm}{\hbar^2}r_0\frac{\hbar^2}{m_e e^2}} = \sqrt{\frac{T}{e^2/r_0}\frac{m}{m_e}} \simeq 5. \tag{4.70}$$

Here, m is the mass of the group of atoms held in position by the spring k, and we have used $T \simeq 25\,\text{meV}$, $e^2/2r_0 \simeq 10\,\text{eV}$, $m \simeq 12\,\text{GeV}$ (for ^{12}C), $m_e \simeq 0.5\,\text{MeV}$. Of course this entropy is even larger for larger m.

4.8 Models Based on Chemical Kinetics (Fisher and Kolomeisky, 1999)

Another approach used to describe, for example, molecular motors and, more generally, conformational transitions, is based on rate equations. The continuum mechanics of the last sections is replaced by transition rates between discrete states. Following Fisher and Kolomeisky, consider kinesin walking along the microtubule (the "track"); let $E_\ell(s)$ represent the enzyme in state s and at the discrete position ℓ along the track, with ℓ an integer (the actual position x_ℓ is such that $x_{\ell+1} - x_\ell = d$, with d the step size of the motor). For example, if we consider three states, a reasonable assignment for

$s = 0, 1, 2$ would be $E, E \cdot ATP, E \cdot ADP \cdot P_i$, where E is the enzyme. Then the enzymatic cycle is described by the six transition rates

$$E_\ell(0) \underset{w_1}{\overset{u_1}{\rightleftharpoons}} E_\ell(1) \underset{w_2}{\overset{u_2}{\rightleftharpoons}} E_\ell(2) \underset{w_3}{\overset{u_3}{\rightleftharpoons}} E_{\ell+1}(0). \tag{4.71}$$

The position of the enzyme is $x = \ell d$, but in the course of time, and for the individual motor, (4.71) is a stochastic process, so we must consider ensemble averages. For example, the average velocity v is given by

$$\langle x \rangle = vt \tag{4.72}$$

(the motor starting from $x = 0$ at $t = 0$). Similarly, one can define a diffusion constant D, describing the width of the probability distribution of x around $\langle x \rangle$, through

$$\left\langle (x - \langle x \rangle)^2 \right\rangle = 2Dt \tag{4.73}$$

(at long times; the factor 2 is because we are in 1-D) One can derive exact expressions for v and D in terms of the rates in (4.71), for any number of states s one may want to consider. With respect to the viscoelastic model of section 4.5, which deals with ensemble-averaged trajectories, we perceive that the advantage of the more microscopic formulation (4.71) is that it leads naturally to consider fluctuations, such as (4.73), that can actually be measured in experiments. On the other hand, kinetic models of the form (4.71) are rarely very predictive, especially as they become more elaborate by adding more states, due to the large number of parameters. One basically shifts the problem to the question of calculating the rates. Nonetheless, it is instructive to follow this viewpoint. Let us therefore consider an even simpler scheme: a two-states motor. Now $E_\ell(0)$ represents the enzyme at site ℓ, and $E_\ell(1)$ is the enzyme with ATP bound. The kinetic scheme is

$$\cdots \underset{w_2}{\overset{u_2}{\rightleftharpoons}} E_\ell(0) \underset{w_1}{\overset{u_1}{\rightleftharpoons}} E_\ell(1) \underset{w_2}{\overset{u_2}{\rightleftharpoons}} E_{\ell+1}(0) \underset{w_1}{\overset{u_1}{\rightleftharpoons}} \cdots . \tag{4.74}$$

To calculate the speed of the motor, consider a steady state situation where the motor is in state 1 with probability p and in state 0 with probability q, and $p + q = 1$. From (4.74), the "motor current" to the right is

$$j = u_2 p - w_2 q = u_1 q - w_1 p, \tag{4.75}$$

from which we get

$$p = \frac{u_1 + w_2}{u_1 + u_2 + w_1 + w_2}, \qquad q = \frac{u_2 + w_1}{u_1 + u_2 + w_1 + w_2}, \tag{4.76}$$

and from (4.75),

$$j = \frac{u_1 u_2 - w_1 w_2}{u_1 + u_2 + w_1 + w_2}. \tag{4.77}$$

This current is the rate for the process $\ell \to \ell + 1$; the speed of the motor is

$$v = jd = \frac{u_1 u_2 - w_1 w_2}{u_1 + u_2 + w_1 + w_2} d. \tag{4.78}$$

If u, w are the zero-load rates, (4.78) gives the zero-load velocity. More generally, we can obtain the force–velocity curve from (4.78) by assuming a dependence of the rates on the force. Since conformational changes are involved, we may assume, within the barrier-crossing scenario,

$$\begin{cases} u_i(f) = u_i(0)e^{-\theta_i^+ fd/T}, \\ w_i(f) = w_i(0)e^{+\theta_i^- fd/T}, \end{cases} \tag{4.79}$$

where f is the load force and the θ_i's are factors ($|\theta_i^\pm| \le 1$) describing, for each transition, the extent of conformational motion in the direction of the force, scaled by d. At this level of description there is no a priori requirement on the values of the θ_i's, except for the overall constraint

$$\theta := \sum_i (\theta_i^+ + \theta_i^-) > 0, \tag{4.80}$$

because the load force f opposes the motion. Now we can insert (4.79) in (4.78) and get an expression for the force–velocity curve $v = v(f)$. Even for a two-steps motor, it depends on many parameters; depending on the choice of $\theta_i^\pm$'s, the graph of v vs. f can be approximately linear, concave, convex, or even non-monotonic! For this reason, the shape of the force–velocity curve is not very informative with respect to the rate models.

Let us now examine the stall force f_{stall}. It can be obtained with an argument very similar to Einstein's original derivation of the Einstein relation. Namely, we imagine adding an insurmountable barrier on the track. The motor on average pushes against the barrier with a force f_{stall}. If x is the position on the track, counted as distance to the barrier, we will see below that the stationary probability for the motor's position is exponential:

$$p(x) \sim e^{-\kappa x/d}. \tag{4.81}$$

With the same legitimacy as Einstein's argument, we associate this probability with a potential energy $\phi(x)$:

$$p(x) \sim e^{-\phi(x)/T}, \tag{4.82}$$

where

$$\phi(x) = f_{\text{stall}} x. \tag{4.83}$$

From (4.81), (4.82), and (4.83) we obtain the stall force:

$$f_{\text{stall}} = \frac{T}{d} \kappa. \tag{4.84}$$

To calculate κ in terms of the rates, and to verify the relation (4.81), we refer to the kinetic scheme (4.74), consider a steady state situation with the motor pushing against the barrier at $\ell = 0$, and employ the notation $p(\ell) = $ probability of the motor being in state 1 at position ℓ and $q(\ell) = $ probability of the motor being in state 0 at position ℓ. The motor occupies the space $\ell \le 0$. We write that the total probability current out of state 1 at ℓ (i.e., $E_\ell(1)$ in (4.74)) is zero, and the same for state 0 (i.e., $E_{\ell+1}(0)$ in (4.74)):

$$\begin{cases} 0 = \frac{\partial}{\partial t}p(\ell) = u_1 q(\ell) + w_2 q(\ell + 1) - (u_2 + w_1)p(\ell), \\ 0 = \frac{\partial}{\partial t}q(\ell + 1) = u_2 p(\ell) + w_1 p(\ell + 1) - (u_1 + w_2)q(\ell + 1). \end{cases} \tag{4.85}$$

Rearranging and changing the index $\ell \to \ell + 1$,

$$\begin{cases} u_2 p(\ell - 1) = (u_1 + w_2)q(\ell) - w_1 p(\ell), \\ u_1 q(\ell - 1) = (u_2 + w_1)p(\ell - 1) - w_2 q(\ell). \end{cases} \tag{4.86}$$

Given $p(\ell), q(\ell)$, the finite differences eqs. (4.86) allow us to calculate $p(\ell - 1)$, $q(\ell - 1)$ and so solve the problem recursively. Keeping in mind our boundary conditions ($p(\ell = 0) = A$, $q(\ell = 0) = B$, where A, B are some nonzero values that we do not need to specify), while $p, q \to 0$ for $\ell \to -\infty$, we make the ansatz

$$p(\ell) = Ae^{\kappa \ell}, \qquad q(\ell) = Be^{\kappa \ell}. \tag{4.87}$$

Substituting into (4.86) we obtain a linear system of equations for A and B,

$$\begin{cases} (u_2 e^{-\kappa} + w_1)A - (u_1 + w_2)B = 0, \\ (u_2 + w_1)e^{-\kappa}A - (u_1 e^{-\kappa} + w_2)B = 0, \end{cases} \tag{4.88}$$

which may have nonzero solutions if

$$\det \begin{pmatrix} (u_2 e^{-\kappa} + w_1) & -(u_1 + w_2) \\ (u_2 + w_1)e^{-\kappa} & -(u_1 e^{-\kappa} + w_2) \end{pmatrix} = 0. \tag{4.89}$$

Equation (4.89) is a quadratic equation for $e^{-\kappa}$ with the two solutions $e^\kappa = 1 \Rightarrow \kappa = 0$, which does not satisfy the boundary conditions, and

$$e^\kappa = \frac{u_1 u_2}{w_1 w_2} \Rightarrow \kappa = \ln \frac{u_1 u_2}{w_1 w_2}. \tag{4.90}$$

In conclusion, the exponential distribution (4.87) solves the stationary problem with κ given by (4.90). From (4.84), the stall force is

$$f_{\text{stall}} = \frac{T}{d} \ln \left(\frac{u_1 u_2}{w_1 w_2} \right), \tag{4.91}$$

where u, w are the zero-load rates. With respect to real motors, this is, of course, an estimate. If one knew, within the model (4.74), (4.79), the values of the many parameters, another estimate for f_{stall} can be derived from the force–velocity curve obtained from (4.79), (4.78). It is an interesting question whether a general relation can be found between the stall force and a set of opportune zero-load parameters of the motor.

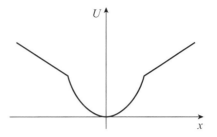

FIGURE 4.10. Potential energy function for the FRJ dynamical system model.

This is an open question; neither (4.91) nor (4.55) nor any other present model has thermodynamic (meaning general) validity.

4.9 Different Levels of Microscopic Description

In the previous sections we took the "thermodynamic" approach of writing equations for the ensemble-averaged trajectories of the system. The goal of statistical mechanics is, of course, to start from the particles and their interactions, which in this case would be the atoms constituting the enzyme, and calculate probability distributions and ensemble averages. For a heterogeneous system with many components, such as the enzyme, this microscopic description must be approached by MD simulations. However, other, intermediate levels of description may also be useful. As an example, we briefly discuss the Fogle–Rudnick–Jasnow (FRJ) heuristic model for a dynamical viscoelastic transition. It is a very simple "energy landscape" model. To set the stage, we imagine grabbing the enzyme at two locations on its surface, of relative position $x(t)$, and exert forces $f(t)$. This situation corresponds to the nano-rheology experiments and, for the statics, to the experiments with the DNA spring. The "microscopic model" for the degree of freedom $x(t)$ consists of a potential energy function $U(x)$ that reflects the existence of a softening transition in the mechanics beyond linear elasticity (figure 4.10). We represent the dynamics of the degree of freedom x as a massless "particle" moving in the potential U, subject to the random force of thermal fluctuations and a corresponding dissipation. If we include thermal noise, we must include dissipation also, because the two phenomena have the same microscopic origin, as expressed in the fluctuation–dissipation theorem. This is the Langevin dynamics approach. The particle is massless because we will be exploring only low frequency dynamics, where inertia is negligible. Also, we may of course add an external force $f(t)$. The potential energy $U(x)$ is really a free energy and depends on temperature: $U = U(x, T)$.

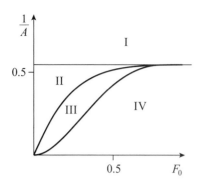

Now, the spirit of the FRJ model is that rather than thinking of $U(x)$ as an energy landscape that somehow mirrors the enzyme structure, it seeks simply to define a minimal dynamical system that exhibits a viscoelastic transition. The focus is then on the characteristics of this transition. This reflects the traditional approach of condensed matter physics, where phase transitions come in certain universality classes, which can therefore be explored through simplified model systems. Their dynamical system is a driven, damped, massless, nonlinear spring. Here, $U(x)$ is the spring's potential energy, and its nonlinearity (figure 4.10) reflects a softening of the system beyond the parabolic linear elasticity regime. Explicitly,

$$U(x) = \begin{cases} \frac{1}{2}\kappa x^2 & \text{for } |x| \leq 1, \\ \frac{\kappa}{2} + F_0|x| & \text{for } |x| > 1, \end{cases} \tag{4.92}$$

that is, a harmonic spring for small strain and a constant restoring force F_0 beyond a yield point $|x| = 1$ (the coordinate x is assumed appropriately rescaled). The FRJ dynamical system is

$$\frac{dx(t)}{dt} = \frac{1}{\gamma}\left\{-\frac{\partial}{\partial x}U[x(t)] + A\sin(\omega t)\right\}, \tag{4.93}$$

where A is the amplitude of the external sinusoidal drive. This deterministic system turns out to be quite interesting in its own right, as it exhibits a surprising complexity of dynamical phase transitions. Figure 4.11 is a sketch of the phase diagram in the regime $\omega < \omega_c \equiv \kappa/\gamma$ (κ is the curvature of the parabolic potential (4.92)). In region I, the steady state, periodic solutions of (4.93), (4.92) are nearly sinusoidal and nearly in phase with the drive (i.e., essentially non-dissipative); in region II, the solutions are distorted (non-sinusoidal), out of phase with the drive (i.e., dissipative), and much larger in amplitude. This is shown in the time traces of figure 4.12, where the solid line corresponds to parameter values just inside region I (namely, $1/A = 0.53$), while for the dashed line we are just inside region IV ($1/A = 0.51$); $F_0 = 0.6$ for both traces.

These solutions are symmetric about $x = 0$ in the sense that the particle spends the same amount of time in the half-space $x > 0$ as in the half-space

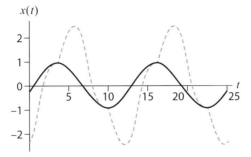

FIGURE 4.12. Steady state solutions for the FRJ dynamical system (4.92), (4.93): just inside region I of the phase diagram ($1/A = 0.53$, solid line), and just inside region IV ($1/A = 0.51$, dashed line); $F_0 = 0.6$ for both cases. Adapted from Fogle, Rudnick, and Jasnow (2015).

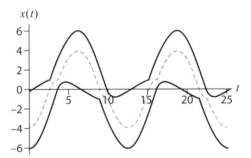

FIGURE 4.13. A pair of steady state, asymmetric solutions for the FRJ dynamical system in region II of the phase diagram ($1/A = 0.45$, $F_0 = 0.1$). The dashed line is a dynamically unstable symmetric solution. Adapted from Fogle, Rudnick, and Jasnow (2015).

$x < 0$. Specifically, these solutions are symmetric under the transformation $x \to -x$, $t \to t + \pi/\omega$, that is, they obey $x(t) = -x(t + \pi/\omega)$: the same symmetry as a sine wave. In regions II and III, in contrast, the stable, steady state solutions are skewed, with the particle, for example, spending most of the time in the region $x > 0$. An example of time traces is given in figure 4.13, which also shows that these symmetry-breaking solutions come in pairs. For each solution with $x > 0$ on average, there is a corresponding solution with $x < 0$ on average. The reason is the symmetry of the equation of motion. Given (4.92), the term $-\partial U/\partial x \equiv f(x)$ in (4.93) has the symmetry $f(-x) = -f(x)$, while $\sin(\omega t + \pi) = -\sin(\omega t)$. Therefore, the transformation

$t \to t + \pi/\omega$ in (4.93) leads to

$$\frac{d}{dt}x(t + \pi/\omega) = \frac{1}{\gamma}\{-f[-x(t + \pi/\omega)] - A\sin(\omega t)\}, \qquad (4.94)$$

and writing $dx/dt = -d(-x)/dt$ we see that $-x(t + \pi/\omega)$ satisfies the same equation as $x(t)$.

Focusing now on the nature of the transitions between the different regions in the phase diagram, one can define a response amplitude I by

$$I^2 = \frac{\omega}{2\pi} \int_0^{2\pi/\omega} x^2(t)\,dt, \qquad (4.95)$$

which has the advantage that it can be unambiguously decomposed into a dissipative part I_d and non-dissipative ("elastic") part I_s: $I^2 = I_d^2 + I_s^2$, with

$$I_d = \sqrt{2}\,\frac{\omega}{2\pi} \int_0^{2\pi/\omega} x(t)\cos(\omega t)\,dt. \qquad (4.96)$$

The right-hand side of (4.96) projects the solution $x(t)$ onto the out-of-phase vector $\cos(\omega t)$; it is referred to the sine wave drive in (4.93). What remains, which is *not* equal to

$$\sim \int_0^{2\pi/\omega} x(t)\sin(\omega t)\,dt \qquad (4.97)$$

because this is a nonlinear system, is non-dissipative. Namely, writing

$$x(t) = \sqrt{2}\,I_d\cos(\omega t) + \left[x(t) - \sqrt{2}\,I_d\cos(\omega t)\right] \equiv x_{\text{diss}}(t) + x_{\text{el}}(t), \qquad (4.98)$$

one has that the dissipation (force times velocity integrated over a cycle) for the component x_{el} is zero:

$$\int_0^{2\pi/\omega} \dot{x}_{\text{el}}(t)\sin(\omega t)\,dt = 0, \qquad (4.99)$$

that is, the dissipation part of the response is entirely contained in $x_{\text{diss}}(t)$ in (4.98).

The response amplitude I (and its dissipative and non-dissipative components) provides an insightful characterization of the transitions. It is found that these are discontinuous ("first order") if the softening transition in the potential is abrupt, as in (4.92). For example, I as a function of the drive amplitude A has a discontinuity when crossing the phase line between regions I and II in figure 4.11. On the other hand, if the softening transition in the potential is smooth, the discontinuity in I disappears and the transition becomes continuous.

To summarize, a nonlinear spring where the nonlinearity consists of a softening transition as in (4.92) gives rise to a very interesting dynamical system exhibiting dynamical phase transitions reminiscent of the sharp viscoelastic transition observed in the nano-rheology experiments on enzymes.

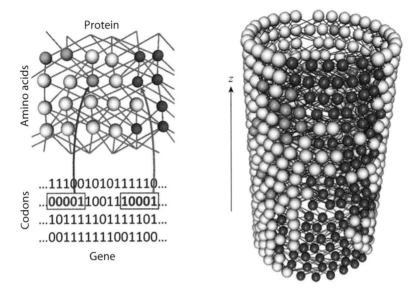

FIGURE 4.14. Mapping from gene to structure in the Tlusty–Libchaber–Eckmann (TLE) model of information flow for allosteric dynamics in enzymes. Adapted from Tlusty, Libchaber, and Eckmann (2017).

4.10 Connection to Information Flow

Like all biological structures, enzymes are engineered by evolution. "Understanding" enzymes means not only understanding the materials properties of the molecule and its dynamics, but also understanding the information flow that created it, and ultimately the true information content of the structure. For an engineered structure, one cannot divorce the properties from the design. Large conformational motion in enzymes often corresponds to a connected region of high strain inside the molecule, such as a shear plane. Tsvi Tlusty and collaborators address the information content of such structures through the following model. Visualize the folded polypeptide chain as a string of beads (amino acids) wrapped in several turns around the surface of a cylinder (figure 4.14).

The bottom of the cylinder represents a patch of surface of the enzyme (say the "lower hemisphere"), the top another patch of surface ("upper hemisphere"); in between is the interior of the enzyme, where each amino acid interacts with the five nearest neighbors (NNs) immediately below it. The interaction is either on (connected) or off (disconnected). In sequence space, the structure is coded for by five-digit binary codons where each "1" corresponds to a connection to the corresponding NN (and "0" means no connection). The relation between structure and mechanics is defined by the following rules. In a given structure, we define two kinds of amino acids with respect to mechanics: rigid (light gray) and fluid shearable (dark gray). Rigid applies to those that connect to at least two rigid neighbors in the row below. Fluid shearable describes those that are not rigid and for which at least one of the three NNs below is fluid shearable. One perceives that there is then also a third mechanical type (intermediate gray): call fluid non-shearable, for example, a disconnected amino acid (codon 00000) surrounded by rigid ones. In the model, a connected fluid-shearable region extending from bottom to top represents a shear plane allowing large conformational motion and, more specifically, an allosteric connection between the bottom and top surfaces.

In terms of information flow, one starts with a prescribed fluidity pattern at the bottom surface, and a random gene. Evolutionary pressure is represented by a target fluidity pattern for the top surface, and the following evolutionary process: a random one-digit mutation of the gene is kept for the next generation except if it is deleterious with respect to the target fluidity pattern of the top. Conceptually, we are asking the system (the protein plus the evolutionary algorithm) to solve the following problem: given a "soft binding site" (of the induced-fit type) at the bottom, find a structure that mechanically connects it to a prescribed soft binding site at the top. For a real enzyme there are of course additional constraints (e.g., the global stability of the structure), but the spirit here is to capture the essential information flow from gene to structure that may give rise to an allosteric enzyme.

Tlusty and collaborators consider a protein of ~ 500 amino acids, corresponding to a gene length of $5 \times 500 = 2500$ "bases." The number 2500 is also the dimension of gene space (the number of coordinates, which are binary valued), while the total number of elements of gene space (and also configuration space, since we have a one-to-one mapping) is $2^{2500} \approx 10^{753}$. Such large numbers show up in the realm of games, for the trivial reason that even a relatively small number of consecutive decisions live on a tree with many branches. For chess, an estimate could be $\sim 10^{80}$ branches (there are $10 \times 10 = 10^2$ possibilities for each pair of moves and a typical game of 40 moves gives a tree with $10^{2 \times 40}$ branches). However, the TLE protein takes typically $\sim 10^4$ moves to win the game, that is, to find a solution as in figure 4.15.

For a given input–output problem, they explored $\sim 10^6$ random initial conditions giving rise to a similar number of different solutions. To characterize the statistics of these solutions, one can look at the dimension of the space of solutions. The dimension is estimated by the box counting algorithm: given an appropriate measure of distance λ between solutions, one counts the number of pairs of solutions differing by less than λ: $N(\lambda)$; the slope of the graph of $\log N(\lambda)$ vs. $\log \lambda$ in the scaling region gives the dimension D. The result is that, in sequence space, the dimension of the set of solutions is still large: $D \geq 150$. But in configuration space, the dimension is dramatically reduced: $D \approx 9$. In words, the genes appear random while the structures are not.

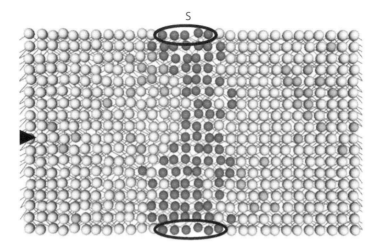

S

FIGURE 4.15. One solution of the TLE model that evolved from a random gene. Adapted from Tlusty, Libchaber, and Eckmann (2017).

The physical meaning of the configurations that solve the input–output problem is that along the channel of fluid-shearable residues connecting the bottom and top binding sites, the structure is easily shearable, with the part on one side of the divide moving essentially like a rigid body with respect to the other side. This kind of motion, involving strain localized to a shear plane or a "hinge" region, is indeed observed among induced-fit-type enzymes, by comparing their X-ray structures with and without ligands. One may also ask whether the linear elasticity response of the structure bears some resemblance to these large amplitude nonlinear deformations. The answer is yes. The situation is somewhat analogous to hydrodynamic instabilities such as viscous fingering, Rayleigh–Bénard convection, and many other examples where linear instability determines one parameter—the most unstable wavelength—which is reflected to some extent in the subsequent nonlinear pattern, even though the dynamics of the large amplitude nonlinear solutions is very different from the dynamics at onset of the instability. In the next section we address the linear elasticity analysis of a structure such as figure 4.15, and, following the TLE model, briefly touch upon the extension of normal mode analysis used to describe the *statistical* properties of the solutions giving rise to allosteric channels as in figure 4.15.

4.11 Normal Mode Analysis

Small elastic deformations of an object with an irregular shape, such as the enzyme of figure 4.7, could be analyzed by solving the equations of continuum elasticity numerically, which implies replacing the continuum volume of the object with a discretized lattice. A similar but simpler approach is to replace the enzyme with a network of point masses connected by springs. The masses replace the individual atoms or groups of atoms in the structure; the springs connect each mass with its nearest neighbors suitably defined. Evidently, different degrees of coarse graining can be implemented. These models are referred to as Gaussian network models. We now recall the notion of normal modes. In general, for a system of masses bound together in a solid state we have a potential energy $U(x_1, x_2, \ldots, x_N)$ and a kinetic energy $\sum_{i=1}^{N} \frac{1}{2} m_i \dot{x}_i^2$, and equilibrium positions $\vec{x}^0 = (x_1^0, x_2^0, \ldots, x_N^0)$; the x_i are coordinates of the masses. Expanding U around the equilibrium position, in terms of the deviation from equilibrium $q_i \equiv x_i - x_i^0$ (which are small quantities), we have

$$U(q_1, q_2, \ldots, q_N) = U(\vec{0}) + \frac{1}{2} \sum_{i,j} \frac{\partial^2 U}{\partial q_i \partial q_j}\bigg|_{\vec{q}=\vec{0}} q_i q_j + \cdots, \qquad (4.100)$$

and since $\dot{x}_i = \dot{q}_i$ the Lagrangian of the system to this order is

$$L(q, \dot{q}) = \sum_i \frac{1}{2} m_i \dot{q}_i^2 - \frac{1}{2} \sum_{i,j} k_{ij} q_i q_j, \qquad (4.101)$$

where

$$k_{ij} = \frac{\partial^2 U}{\partial q_i \partial q_j}\bigg|_{\vec{q}=\vec{0}} = \frac{\partial^2 U}{\partial x_i \partial x_j}\bigg|_{\vec{x}=\vec{x}^0}. \qquad (4.102)$$

This is the Gaussian or harmonic approximation. The equations of motion are obtained from

$$\frac{d}{dt} \frac{\partial L}{\partial \dot{q}_i} - \frac{\partial L}{\partial q_i} = 0 \qquad (4.103)$$

$$: \quad m_i \ddot{q}_i + \frac{1}{2} \sum_j k_{ij} q_j = 0. \qquad (4.104)$$

For illustration purposes, we take the simplest situation: two masses in 1-D connected by a spring. Then

$$U(x_1, x_2) = \frac{1}{2} k (x_2 - x_1 - a)^2, \qquad (4.105)$$

where x_1, x_2 are the positions of the masses m_1, m_2, and a is the equilibrium distance between them:

$$0 = \frac{\partial U}{\partial x_1} = \frac{\partial U}{\partial x_2} \Rightarrow x_2 - x_1 = a. \qquad (4.106)$$

Taking coordinates $q_1 = x_1$, $q_2 = x_2 - a$ the Lagrangian is

$$L = \frac{1}{2}(m_1 \dot{q}_1^2 + m_2 \dot{q}_2^2) - \frac{1}{2} k (q_1 - q_2)^2, \qquad (4.107)$$

and one obtains the equations of motion

$$\begin{cases} m_1\ddot{q}_1 - k(q_2 - q_1) = 0, \\ m_2\ddot{q}_2 + k(q_2 - q_1) = 0, \end{cases} \tag{4.108}$$

which we can write in matrix form

$$\frac{d^2}{dt^2} M \begin{pmatrix} q_1 \\ q_2 \end{pmatrix} + K \begin{pmatrix} q_1 \\ q_2 \end{pmatrix} = \vec{0} \tag{4.109}$$

or

$$\frac{d^2}{dt^2} M_{ik}q_k + K_{ik}q_k = 0, \tag{4.110}$$

where

$$M = \begin{pmatrix} m_1 & 0 \\ 0 & m_2 \end{pmatrix}, \quad K = \begin{pmatrix} k & -k \\ -k & k \end{pmatrix}. \tag{4.111}$$

With the ansatz

$$q_k = A_k e^{i\omega t}, \tag{4.112}$$

eq. (4.109) becomes

$$\tilde{M} \begin{pmatrix} A_1 \\ A_2 \end{pmatrix} = \vec{0}, \tag{4.113}$$

where

$$\tilde{M} = \begin{pmatrix} -\omega^2 m_1 + k & -k \\ -k & -\omega^2 m_2 + k \end{pmatrix}. \tag{4.114}$$

From (4.113) we see that the eigenfrequencies of the problem are found from

$$\det \tilde{M} = 0. \tag{4.115}$$

The corresponding solutions of (4.113), that is, the eigenvectors (A_1, A_2), are the normal modes, which perform simple harmonic oscillations according to (4.112).

For the present case, (4.115) gives

$$m_1 m_2 \omega^4 - k(m_1 + m_2)\omega^2 = 0 \tag{4.116}$$

$$\Rightarrow \quad \omega = 0 \quad \text{or} \quad \omega = \sqrt{\frac{k}{\mu}}, \quad \mu = \frac{m_1 m_2}{m_1 + m_2}. \tag{4.117}$$

For the zero eigenvalue, from (4.113) we find

$$A_1 = A_2, \quad \text{i.e., } q_1 = q_2 \Rightarrow x_2 - x_1 = a. \tag{4.118}$$

This is a translational mode (uniform translation of the center of mass of the system, with no oscillation). For the nonzero eigenvalue we find

$$A_2 = -\frac{m_1}{m_2}A_1, \quad \text{i.e.,} \quad \begin{cases} x_1(t) = A\cos(\omega t), \\ x_2(t) = -A\frac{m_1}{m_2}\cos(\omega t). \end{cases} \tag{4.119}$$

This is a simple harmonic oscillation of the coordinate $(\frac{1}{2}x_1 - \frac{1}{2}\frac{m_2}{m_1}x_2)$, with the center of mass at rest.

Coming back to the general equation (4.104), for the special case that the masses are all the same ($m_i = m$ for all i), with the same ansatz (4.112) we find

$$K_{ij}A_j = m\omega^2 A_i \quad (K_{ij} = \tfrac{1}{2}k_{ij}). \tag{4.120}$$

The algorithm is then to find the eigenvalues λ_i of the matrix K: the eigenfrequencies are given by $\lambda_i = m\omega_i$ and the corresponding eigenvectors $\vec{A}_i$ represent the normal modes.

For a spring network representing the structure in figure 4.7, from this analysis one would find that one of the low frequency normal modes, when represented with a "large" amplitude (much beyond the physical linear-elasticity regime of the structure), corresponds to a deformation roughly similar to the conformational transition shown in the figure. Thus the linear elasticity analysis gives an indication of the kind of large amplitude deformations of the structure that may occur for the real enzyme. However, the dynamics of the actual conformational motion is completely different. The smallest eigenfrequencies one obtains using realistic values for spring constants and masses are many orders of magnitude larger than the actual rates of large conformational motion. We can see this by considering a cube of elastic material of size $L \sim 5\,\text{nm}$ (the size of the enzyme). Elastic waves in this object obey the wave equation

$$\frac{\partial^2 u}{\partial x^2} - \frac{1}{c^2}\frac{\partial^2 u}{\partial t^2} = 0, \quad c \sim \sqrt{\frac{Y}{\rho}}, \tag{4.121}$$

where $u(x, t)$ is the displacement field, and we do not distinguish longitudinal and transverse waves for this order of magnitude argument; Y is Young's modulus and ρ is the density; c is the speed of sound. Considering a plane wave solution of (4.121),

$$u(x, t) = A\sin(kx - \omega t), \quad \omega = ck, \tag{4.122}$$

the fundamental mode corresponds to the longest wavelength:

$$\lambda_{\max} = 2L \Rightarrow \omega_{\min} = \frac{2\pi c}{\lambda_{\max}} = \pi\frac{c}{L}. \tag{4.123}$$

Using for Young's modulus a lower-end value,

$$Y \sim 10\,\text{MPa} = 10^8\,\text{dyn/cm}^2, \quad \rho \sim 1\,\text{g/cm}^3, \quad L = 5\,\text{nm}, \tag{4.124}$$

we find $c \sim 10^4$ cm/s and $\omega_{min} \sim 10^{12}$ Hz. Normal mode frequencies of even the "soft modes" are in the THz region (corresponding to the far infrared of the electromagnetic spectrum), while rates for large conformational motion are of order $\sim 10^3$ Hz.

With the structure of figure 4.15, a similar normal mode analysis of any one specific solution would return soft modes corresponding to shear motion along the fluid-shearable channel. There are, of course, many such solutions; Tlusty et al. explore $\sim 10^6$ of them. The question arises of how to characterize the set of solutions statistically. Beyond the dimension of the set, discussed before, it is interesting to deploy a generalization of the normal mode analysis of this section, called singular value decomposition (SVD). Without detailing the math, the general idea as applied, for example, to the set of solutions in sequence space is as follows.

A solution is defined by a string of 10^3 "bases" (we use round numbers here), a base being either 0 or 1. The set of 10^6 solutions is thus represented by 10^6 binary vectors with 10^3 components. The $10^6 \times 10^3$ rectangular matrix formed with the components of these vectors can be "diagonalized," yielding eigenvalues and eigenvectors that describe the structure of the correlations in the set. We leave the reader to play with the basic observation that given a rectangular $m \times n$ matrix A $(m > n)$, the matrix $A^t A$ is a square $n \times n$ matrix. The result of this analysis for the set of solutions both in sequence space and configuration space is that there are 8–9 isolated "soft modes" apart from the near-continuum expressing a near-random distribution. This number corresponds to the dimension of the space of solutions.

4.12 Many States of the Folded Protein: Spectroscopy

The deformability of enzymes corresponds, in atomic structural terms, to the availability of many different conformational states for the folded structure of the protein. We are not talking about elastic modes here: these are common to all solids, and one does not refer to different conformational states corresponding to the presence of a sound wave in the solid. In contrast, the relation between conformational states and deformability of enzymes is similar to the relation between the presence of defects and plastic deformations in macroscopic crystalline solids. Different distributions of defects could be called different "conformations" of the solid, and it is the motion of these defect lines and walls (i.e., the switching between different "conformations") that controls the dynamics of plastic deformations. The notion that the folded protein can explore many different conformational states was dramatically demonstrated in the seventies, experimentally by the group of Frauenfelder, and numerically by the MD simulations of Karplus and collaborators.

In the experiments, the time course of rebinding of oxygen (and also carbon monoxide) to the heme group of myoglobin (Mb) after flash photolysis is followed spectroscopically (most simply by absorption in the blue region of the visible spectrum) in a wide temperature range, from 40 K to room temperature. There are several steps in the rebinding process, but in the temperature region 40–200 K one barrier-crossing process dominates. Here the sample is in a 3:1 glycerol–water mixture, and below the glass transition temperature of the solvent. The key observation is that the rebinding kinetics is non-exponential (and rather close to a power law at any given temperature). At higher temperatures (above 200 K), the same process shows regular exponential kinetics. The non-exponential kinetics of rebinding is interpreted in terms of a distribution of barrier heights. Different conformations of the protein give rise to different barrier heights seen by the ligand in the process of rebinding to the heme. At low temperature, each protein molecule is frozen in a particular conformation, so the sample is effectively heterogeneous. For a given barrier height ε, the rate k for the ligand to move across it is given by the Arrhenius factor

$$k = k_0 e^{-\varepsilon/T}. \tag{4.125}$$

In the course of time, the fraction of Mb molecules that have not rebound to a ligand (i.e., the probability that an Mb molecule is ligand-free at time t after flash photolysis at time 0) is

$$f_1(t, k) = e^{-kt}. \tag{4.126}$$

With a "frozen" distribution of barriers $g(\varepsilon)$ (where $g(\varepsilon)\, d\varepsilon$ is the probability that the barrier height is between ε and $\varepsilon + d\varepsilon$), the fraction of ligand-free molecules becomes

$$f(t) = \int_0^\infty d\varepsilon\, g(\varepsilon) f_1(t, k), \tag{4.127}$$

with f_1 given by (4.126) and the relation between k and ε given by (4.125). Changing variables in the integral from ε to k (with $d\varepsilon = -T\, dk/k$) one can write

$$f(t) = T \int_0^{k_0} dk\, \frac{g[\varepsilon(k)]}{k} e^{-kt} \approx T \int_0^\infty dk\, \frac{g[\varepsilon(k)]}{k} e^{-kt}, \tag{4.128}$$

the last approximate equality holding because $k_0 t \gg 1$ throughout the experimental data. Figure 4.16 shows experimental measurements for CO (upper half) and O_2 (lower half) rebinding to myoglobin (Mb) in the temperature range 40–160 K. The graph is log-log, showing that the dynamics follows approximately a power law at any given temperature. For comparison, an exponential function is shown by the dashed line.

One may wonder whether this glassy dynamics is forced by the solvent, which is itself a glass. The understanding is, however, that the glassy

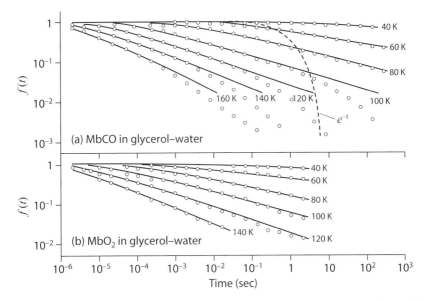

FIGURE 4.16. Spectroscopic measurements (absorption at 436 nm) of the time course of CO and O_2 rebinding to Mb after flash photolysis. The sample is in a glycerol–water mixture which is solid (below the glass transition temperature) at the temperatures shown. The log-log plot shows that the rebinding dynamics is non-exponential, and follows approximate power laws $\sim (1 + t/t_0)^{-n}$ with a temperature-dependent exponent (solid lines). The dashed line shows an exponential function for comparison. Adapted from Austin et al. (1975).

solvent traps and freezes the Mb molecules into different conformations. The observed dynamics is the dynamics of ligand rebinding to this heterogeneous population of Mb molecules, and not the dynamics of conformational motion of the proteins or the surrounding solvent, which are frozen.

The different conformations into which the Mb molecules are trapped are represented by a distribution of barriers for ligand rebinding, $g(\varepsilon)$. From (4.128) we see that the experimentally measured function $f(t)$ is the Laplace transform of the function $Tg[\varepsilon(k)]/k$; the distribution g can therefore be (approximately) obtained from the experimental data at a single temperature T by finding a function whose Laplace transform approximates $f(t)$. The $g(\varepsilon)$ obtained in this way is shown in figure 4.17 for three different cases. Figure 4.18 shows again the measured $f(t)$ for CO rebinding (the same data as in figure 4.16), but the lines are now calculated from (4.127):

$$f(t) = \int_0^\infty d\varepsilon\, g(\varepsilon) f_1(t, k) = \int_0^\infty d\varepsilon\, g(\varepsilon) e^{-k(\varepsilon)t}, \qquad (4.129)$$

where $k = k(\varepsilon)$ is given by (4.125). The result is that a single, temperature-independent barrier distribution $g(\varepsilon)$ accounts for all the different experimental curves for $40 \leq T \leq 160\,\mathrm{K}$.

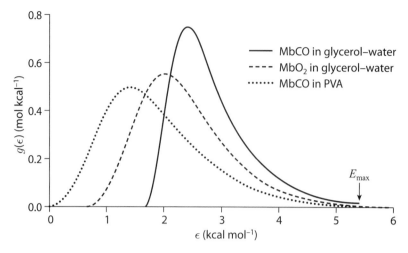

FIGURE 4.17. The experimentally determined barrier distribution $g(\varepsilon)$ for three different cases as indicated. Adapted from Austin et al. (1975).

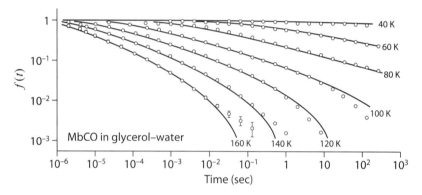

FIGURE 4.18. Rebinding dynamics of CO to Mb. Same data as in figure 4.16, but the lines for the different temperatures are now calculated according to (4.129), using the same barrier distribution $g(\varepsilon)$, shown in figure 4.17. Adapted from Austin et al. (1975).

   In summary, non-exponential ligand rebinding dynamics at low temperature is indicative of a distribution of conformational states for the folded protein. At room temperature, the same rebinding dynamics is in fact exponential. The reason is the fast exchange between different conformational states at room temperature. If the relaxation between different conformational states is faster than any rebinding rate, then the barrier distribution $g(\varepsilon)$ remains the same at all times during the rebinding experiment. This scenario is opposite to the case of "frozen molecules" (no relaxation at all) just discussed, where the barrier distribution among the ligand-free molecules is $g(\varepsilon)$ at time 0, but it changes with time because the sample gets depleted

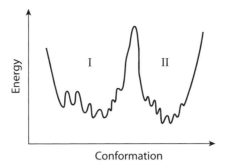

FIGURE 4.19. A rugged energy landscape.

of molecules with low barriers. In the fast relaxation case, the probability of occurrence of a barrier between ε and $\varepsilon + d\varepsilon$ *at time t* is

$$df = f(t)g(\varepsilon)\, d\varepsilon. \tag{4.130}$$

These molecules rebind with rate $k(\varepsilon) = k_0 e^{(-\varepsilon/t)}$, thus

$$\frac{d}{dt}(df) = -k(\varepsilon)f(t)\, d\varepsilon. \tag{4.131}$$

Integrating over ε,

$$\frac{d}{dt}f = -f(t)\int_0^\infty d\varepsilon\, k(\varepsilon)g(\varepsilon) = -\bar{k}f(t), \tag{4.132}$$

so that the dynamics is again exponential,

$$f(t) = e^{-\bar{k}t}, \tag{4.133}$$

with an effective rate

$$\bar{k} = \int_0^\infty d\varepsilon\, k(\varepsilon)g(\varepsilon). \tag{4.134}$$

In the energy landscape picture, the many different conformational states produce a "rugged" or "hierarchical" landscape (figure 4.19), which has the following consequence. The diffusion dynamics of a particle in the potential of figure 4.19, which stands for the diffusive conformational dynamics of the protein, consists of confined diffusion within region I and occasionally a flight into region II, and so on. In terms of rates, there are (at least) two very different ones: a (fast) rate for moving between different states within basin I (or within basin II), and a (slow) rate for moving between basins I and II. Herein lies the connection between the spectroscopic experiments, which underlie the energy landscape picture, and the nano-rheology experiments, which underlie the continuum mechanics description. Namely, a solid that can be characterized by an energy landscape similar to figure 4.19 would in essence be viscoelastic: on short timescales (compared to the Maxwell relaxation time $1/\omega_c$) it deforms elastically within basin I, or basin II, etc. On long timescales, in the presence of an applied stress, it "flows" between different basins. In essence, this is the picture proposed in 2007 by Onuchic and collaborators based on their MD simulations of conformational motion in adenylate kinase. In their terminology, the viscous flow part is called "cracking," and it connects two distant conformations of the enzyme, specifically the open and closed forms. In fact, the concept that the folded protein may be regarded, to some extent, as fluid, was introduced very early on, through the pioneering MD simulations of a whole protein by Karplus and collaborators. However, the story is not closed yet because we do not have, at present, a complete quantitative theory relating to all aspects seen in the experiments and MD simulations. A "standard model" of enzyme conformational dynamics is yet to come.

4.13 Interesting Topics in Nonequilibrium Thermodynamics Relating to Enzyme Dynamics

We discuss the *Langevin equation with moving parameters* (Schmiedl and Seifert, 2007). The setting is one where a thermodynamic system is taken from one equilibrium state to another by changing external thermodynamic parameters (e.g., volume, temperature) according to a specific protocol in time (i.e., the time course of the external parameters is imposed). One can ask what the average dissipation associated with a particular protocol is. "Average" refers to the case of a small system, where fluctuations make the dissipation into a stochastic quantity.

We consider a Brownian particle in 1-D, trapped by a potential $U(x, \lambda)$; x is the coordinate of the particle, λ is a (set of) parameter(s) describing the potential (e.g., strength and location). Neglecting inertia, the Langevin equation for the particle is

$$\dot{x} = -\mu \frac{\partial U}{\partial x} + \Gamma(t), \qquad (4.135)$$

where the random forcing Γ describes equilibrium thermal fluctuations (Γ/μ has dimensions of force),

$$\langle \Gamma(t) \rangle = 0, \qquad \langle \Gamma(t)\Gamma(t') \rangle = 2\mu T \, \delta(t - t'), \qquad (4.136)$$

in that the strength of Γ obeys the fluctuation–dissipation theorem. The stochastic description (4.135) for the position of the particle can be transformed into a deterministic description for the probability distribution $p(x, t)$ of finding the particle at position x at time t. There is a conserved quantity, namely the total probability of finding the particle anywhere:

$$\int_{-\infty}^{+\infty} p(x,t)\,dx = 1 \quad \forall t. \tag{4.137}$$

The conservation law written for the density of this conserved quantity (i.e., p) then reads

$$\frac{\partial p}{\partial t} + \vec{\nabla} \cdot \vec{j} = 0, \tag{4.138}$$

where the current $\vec{j}$ has an advection and a diffusion part:

$$\vec{j} = \langle \dot{x} \rangle\, p(x,t) - D\frac{\partial p}{\partial x}. \tag{4.139}$$

The Einstein relation relates the diffusion coefficient to the mobility, $D = \mu T$, while averaging (4.135) we see that

$$\langle \dot{x} \rangle = -\mu \frac{\partial U}{\partial x}. \tag{4.140}$$

Putting everything together we obtain the Fokker–Planck equation for the probability distribution p:

$$\frac{\partial p(x,t)}{\partial t} - \mu\frac{\partial}{\partial x}\left(\frac{\partial U}{\partial x}p\right) - \mu T\frac{\partial^2 p}{\partial x^2} = 0. \tag{4.141}$$

Now consider a nonequilibrium process (one that is not quasi-static) where the thermodynamic parameter λ, describing the potential, goes from an initial value λ_i at $t = 0$ to a final value λ_f at $t = \tau$, according to a defined protocol $\lambda(t)$. The work done on the system is the integrated power dU/dt:

$$W[\lambda(t)] = \int_0^\tau \left\langle \frac{\partial U}{\partial \lambda}\frac{d\lambda}{dt} \right\rangle dt = \int_0^\tau dt\, \frac{d\lambda}{dt} \left\langle \frac{\partial U}{\partial \lambda} \right\rangle, \tag{4.142}$$

which therefore is a functional of the protocol $\lambda(t)$. The meaning of the ensemble average in (4.142) is provided by the probability distribution described by (4.141), for example,

$$\langle U \rangle(t) = \int_{-\infty}^{+\infty} dx\, U(x)p(x,t)$$

$$\text{and} \quad \left\langle \frac{\partial U}{\partial \lambda} \right\rangle(t) = \int_{-\infty}^{+\infty} dx\, \frac{\partial U(x,\lambda)}{\partial \lambda}p(x,t), \tag{4.143}$$

so that, to calculate the integral (4.142) numerically, we would start from the equilibrium ($\partial p/\partial t = 0$) distribution solution of (4.141) with $U = U(x,\lambda_i)$; at the next time interval we increment λ according to the protocol $\lambda(t)$, calculate the new $p(x)$ using (4.141), and so calculate $\partial U/\partial \lambda$ using (4.143), and so on.

To further illustrate this, take the explicit case of a harmonic trap being displaced according to the protocol $\lambda(t)$:

$$U(x,\lambda) = \frac{1}{2}K[x - \lambda(t)]^2, \tag{4.144}$$

where λ is the position of the minimum of the potential, and we take $\lambda_i = \lambda(0) = 0$, $\lambda_f = \lambda(\tau)$. With the notation $u \equiv \langle x \rangle$, we have from the Langevin eq. (4.135),

$$\langle \dot{x} \rangle = \dot{u} = -\mu \left\langle \frac{\partial U}{\partial x} \right\rangle = -\mu K(\langle x \rangle - \lambda). \tag{4.145}$$

Since $[\mu] = \frac{\text{velocity}}{\text{force}} = \frac{\text{time}}{\text{mass}}$ and $[K] = \frac{\text{force}}{\text{length}}$, but there is no mass in this problem, μ and K appear only as the combination μK; here $1/(\mu K)$ is a characteristic relaxation time. From (4.145) we have the equation for the average position of the particle, $u(t)$:

$$\dot{u} = \mu K(\lambda - u), \tag{4.146}$$

which, given $\lambda(t)$, allows us to calculate $u(t)$. For example, if we move the trap at constant speed v,

$$\lambda(t) = vt, \tag{4.147}$$

eq. (4.146) becomes

$$\dot{u} + \mu K u = \mu K v t. \tag{4.148}$$

The solution with boundary condition $u(0) = 0$ is

$$u(t) = vt + \frac{v}{\mu K} \left(e^{-\mu K t} - 1 \right), \tag{4.149}$$

which shows that for long times ($t \gg 1/(\mu K)$), the average position of the particle, u, lags behind the center of the trap by a distance $v/(\mu K)$.

For the relatively simple case of the displaced trap (4.144), the work can be calculated exactly for any given protocol $\lambda(t)$, by expressing it in terms of $\dot{u}(t)$, which itself is calculated from (4.146). From the expression (4.142) for the work, since $\partial U/\partial \lambda = -K(x - \lambda)$ for the process (4.144), we obtain

$$W = \int_0^\tau dt \, \dot{\lambda} K(\lambda - \langle x \rangle) = \int_0^\tau dt \, \frac{1}{\mu} \dot{\lambda} \dot{u}. \tag{4.150}$$

Using (4.146), we express $\dot{\lambda}$ as

$$\dot{\lambda} = \frac{1}{\mu K} \ddot{u} + \dot{u}, \tag{4.151}$$

and thus

$$W = \int_0^\tau dt \, \frac{1}{\mu} \dot{u} \left(\frac{1}{\mu K} \ddot{u} + \dot{u} \right). \tag{4.152}$$

Integrating the first term by parts, we finally obtain

$$W = \frac{1}{\mu} \int_0^\tau dt \, (\dot{u})^2 + \frac{1}{\mu} \frac{1}{2\mu K} \dot{u}^2 \Big|_0^\tau, \tag{4.153}$$

which is proportional to the inverse mobility. Given a protocol $\lambda(t)$, we can now compute the work using (4.153) and (4.146). For example, for the uniform displacement (4.147), using (4.149) and (4.153) one finds

$$W = \frac{v^2}{\mu} \left[\tau + \frac{1}{\mu K} \left(e^{-\mu K \tau} - 1 \right) \right], \tag{4.154}$$

or, in terms of $\lambda_f = v\tau$,

$$W = \frac{\lambda_f^2}{\mu \tau} \left[1 + \frac{1}{\mu K \tau} \left(e^{-\mu K \tau} - 1 \right) \right]. \tag{4.155}$$

If the protocol is sufficiently slow compared to the characteristic relaxation time of the system, that is, in the limit $\mu K \tau \gg 1$, this is just the hydrodynamic dissipation $W = v^2 \tau / \mu$, where v/μ is the force applied to the particle, $v \times v/\mu$ is the power, and $(v^2/\mu) \times \tau$ is the work. In the opposite limit of a very fast process ($\mu K \tau \ll 1$), we see from (4.155) that $W \approx 0$ for finite λ_f. No work is done, for the same reason that no work is done in the free expansion of a gas.

Contrary to what one might think, the work (4.155) is not the minimal work for processes starting at $\lambda_i = 0$ at $t = 0$ and ending at λ_f at $t = \tau$, that is, the uniform displacement protocol (4.147) does not produce the minimal work. To find the "best" protocol, we look for $u(t)$ that minimizes the right-hand side of (4.153); this expression has the form of an "action" with a "Lagrangian" $L(u, \dot{u}) = \dot{u}^2$. Minimization thus leads to the Euler–Lagrange equation

$$\frac{d}{dt} \frac{\partial L(u, \dot{u})}{\partial \dot{u}} - \frac{\partial L}{\partial u} = 0, \tag{4.156}$$

which in this case gives

$$\ddot{u} = 0 \Rightarrow u = at, \tag{4.157}$$

where we have used the boundary condition $\lambda_i = 0 \Rightarrow u(0) = 0$; the constant a is to be determined. The protocol $\lambda(t)$ corresponding to $u(t)$ given by (4.157) is, using (4.146),

$$\lambda(t) = \frac{1}{\mu K} \dot{u} + u = \frac{a}{\mu K} + at. \tag{4.158}$$

We see that to satisfy the boundary condition $0 = \lambda_i = \lambda(0)$, we need a jump in λ at $t = 0$, that is, $\lambda(0^+) = a/(\mu K)$, while to satisfy the b.c. $\lambda(\tau) = \lambda_f$ we may need another jump in λ at $t = \tau$, since $\lambda(\tau^-) = a/(\mu K) + a\tau$. Figure 4.20 shows the minimum work protocol in terms of λ, u, and $\dot{u}$. To determine a, we minimize the work (4.153):

$$W = \frac{1}{\mu} \int_0^\tau dt\, a^2 + \frac{1}{\mu} \frac{(\mu K)^2}{2\mu K} (\lambda_f - a\tau)^2 = \frac{\tau}{\mu} a^2 + \frac{1}{2} K (\lambda_f - a\tau)^2, \tag{4.159}$$

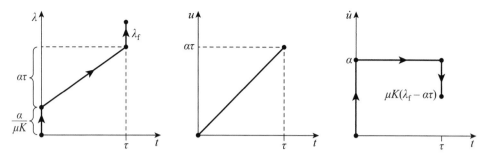

FIGURE 4.20. Minimum work protocol for displacing a harmonic trap that confines a Brownian particle, with λ the position of the trap and u the average position of the particle.

and with respect to a,

$$0 = \frac{\partial W}{\partial a} = 2\frac{\tau}{\mu}a - K\tau(\lambda_f - a\tau) \Rightarrow a = \frac{\mu K}{2 + \mu K\tau}\lambda_f. \tag{4.160}$$

The minimum work protocol is finally

$$\lambda(t) = \frac{\lambda_f}{2 + \mu K\tau}(1 + \mu Kt) \quad \text{for } 0 < t < \tau, \tag{4.161}$$

with jumps of size

$$\Delta = \frac{\lambda_f}{2 + \mu K\tau} \tag{4.162}$$

at the beginning and end, that is,

$$\lambda(0^+) - \lambda_i = \lambda_f - \lambda(\tau^-) = \Delta. \tag{4.163}$$

From (4.159) and (4.160), this minimum work is

$$W_{\min} = \frac{1}{2}K\frac{2}{2 + \mu K\tau}\lambda_f^2. \tag{4.164}$$

For the slow process ($\mu K\tau \gg 1$), $W_{\min}$ again reduces to the value given by the hydrodynamic dissipation:

$$W_{\min} \to \frac{K}{\mu K\tau}\lambda_f^2 = \frac{v^2\tau}{\mu}, \tag{4.165}$$

where $v = \lambda_f/\tau$ is the velocity of the trap, since the jumps go to zero in this limit.

For the fast process ($\mu K\tau \ll 1$), $W_{\min}$ does not go to zero; instead $W_{\min} \to \frac{1}{2}K\lambda_f^2$. The reason is that, in this limit, the minimum work protocol consists of *two* fast jumps of size $\lambda_f/2$ separated by a short waiting time; during the second jump, the external agent has to work against the pressure of the particle.

One can similarly consider a protocol that consists in changing the strength of the trap, that is, $\lambda(t) \to K(t)$: the interested reader should

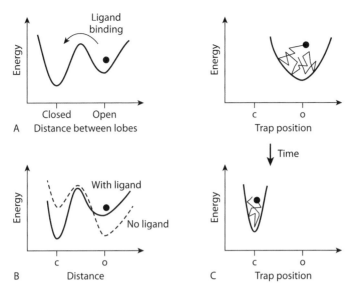

FIGURE 4.21. (A) and (B): energy landscape representations of the induced-fit process of ligand binding. In (A) the conformational transition of the enzyme is represented as a barrier-crossing process for a Brownian particle. In (B) the conformational transition is represented as a change in the energy landscape, such that the absolute minimum corresponds to the open conformation when the ligand is not bound, and to the closed conformation when the ligand is bound. (C) represents the nonequilibrium process of moving a trap from the position "o" (open) to the position "c" (closed) while also changing the strength of the trap. Adapted from Zulkowski et al. (2012).

consult the paper by Schmiedl and Seifert (2007). Furthermore, Zulkowski et al. (2012) solve the case of a more general protocol, where the position of the trap, its strength, and the temperature of the bath are varied at the same time. Remarkably, their solution for the minimum dissipation is based on differential geometry. They calculate the trajectories of minimum dissipation for this system as the geodesics of a manifold defined by a metric tensor constructed with the spring constant K, the dissipation γ, and the inverse temperature $\beta = 1/T$.

The relation of these nonequilibrium processes to enzyme conformational dynamics is both obvious and vague. Figure 4.21 shows one representation of the "induced-fit" process of ligand binding, in terms of a free energy landscape plotted against a conformation coordinate which, in this case, we can specifically associate with some measure of distance between the two lobes of the molecule (we refer to a typical structure as in figure 4.7). The binding event drives the structure from the "open" to the "closed" conformation. Actually, from the viewpoint of the energy landscape, the process might be drawn as in (B): before the binding event, the free energy of the molecule has an absolute minimum for the open conformation; with the ligand bound, the minimum shifts to the closed conformation. The process is then similar to the nonequilibrium process just discussed (C), where the trap position, and possibly the strength, change in time. However, in the description (B) there is no Brownian particle in the trap! Actually, in the case of the enzyme, the "trap" and the "Brownian particle" are the same system, namely the molecule; this introduces logical difficulties in directly relating the processes of figure 4.21 to enzyme dynamics.

BIBLIOGRAPHY

Chapter 1. Brownian Motion

[1] S. Chandrasekhar. Brownian motion, dynamical friction, and stellar dynamics. *Rev. Mod. Phys.*, 21:383, 1949.

[2] A. Einstein. Uber die von der molekularkinetischen Theorie der Warme geforderte Bewegung von in ruhenden Flussigkeiten suspendierten Teilchen. *Ann. Phys.* (in German), 322:549–560, 1905.

[3] A. Einstein. *Investigations on the Theory of the Brownian Movement.* (English translation of the 1905 article.) Dover, 1956.

[4] P.-G. de Gennes. *Introduction to Polymer Dynamics.* Cambridge University Press, 1990.

[5] C. Kittel. *Elementary Statistical Physics.* Dover, 1990.

[6] H. A. Kramers. Brownian motion in a field of force and the diffusion model of chemical reactions. *Physica*, 7:284–304, 1940.

[7] S.-K. Ma. *Statistical Mechanics.* World Scientific, 1985.

[8] H. Quian. Stochastic physics, complex systems and biology. *Quantitative Biology*, 1:50–53, 2013.

[9] F. Reif. *Fundamentals of Statistical and Thermal Physics.* McGraw-Hill, 1965.

[10] K. Sneppen and G. Zocchi. *Physics in Molecular Biology.* Cambridge University Press, 2005.

Chapter 2. Statics of DNA Deformations

[1] S. Ares, N. K. Voulgarakis, K. Ø. Rasmussen, and A. R. Bishop. Bubble nucleation and cooperativity in DNA melting. *Phys. Rev. Lett.*, 94:035504, 2005.

[2] N. B. Becker and R. Everaers. DNA nanomechanics in the nucleosome. *Structure*, 17:579–589, 2009.

[3] C. Bustamante, Z. Bryant, and S. B. Smith. Ten years of tension: Single-molecule DNA mechanics. *Nature*, 421:423–427, 2003.

[4] C. Bustamante, J. F. Marko, E. D. Siggia, and S. Smith. Entropic elasticity of lambda-phage DNA. *Science*, 265:1599–1600, 1994.

[5] T. E. Cloutier and J. Widom. Spontaneous sharp bending of double-stranded DNA. *Mol. Cell*, 14:355–362, 2004.

[6] P. Cluzel, A. Lebrun, C. Heller, R. Lavery, J. L. Viovy, D. Chatenay, and F. Caron. DNA: An extensible molecule. *Science*, 271:792–794, 1996.

[7] F. H. C. Crick and A. Klug. Kinky helix. *Nature*, 255:530–533, 1975.

[8] H. R. Drew, R. M. Wing, T. Takano, C. Broka, S. Tanaka, K. Itakura, and R. E. Dickerson. Structure of a B-DNA dodecamer: Conformation and dynamics. *Proc. Natl. Acad. Sci. USA*, 78:2179–2183, 1981.

[9] Q. Du, A. Kotlyar, and A. Vologodskii. Kinking the double helix by bending deformation. *Nucleic Acids Res.*, 36:1120–1128, 2008.

[10] Q. Du, C. Smith, N. Shiffeldrim, M. Vologodskaia, and A. Vologodskii. Cyclization of short DNA fragments and bending fluctuations of the double helix. *Proc. Natl. Acad. Sci. USA*, 102:5397–5402, 2005.

[11] P. J. Flory. *Principles of Polymer Chemistry*. Cornell University Press, 1953.

[12] S. Geggier, A. Kotlyar, and A. Vologodskii. Temperature dependence of DNA persistence length. *Nucleic Acids Res.*, 39:1419–1426, 2010.

[13] P.-G. de Gennes. *Introduction to Polymer Dynamics*. Cambridge University Press, 1990.

[14] Y. He, M. Maciejczyk, S. Ołdziej, H. A. Scheraga, and A. Liwo. Mean-field interactions between nucleic-acid-base dipoles can drive the formation of the double helix. *Phys. Rev. Lett.*, 110:098101, 2013.

[15] V. Ivanov, D. Piontkovski, and G. Zocchi. Local cooperativity mechanism in the DNA melting transition. *Phys. Rev. E*, 71:041909, 2005.

[16] V. Ivanov, Y. Zeng, and G. Zocchi. Statistical mechanics of base stacking and pairing in DNA melting. *Phys. Rev. E*, 70:051907, 2004.

[17] C. Kittel. Phase transition of a molecular zipper. *Am. J. Phys.*, 37:917–920, 1969.

[18] F. Lankas, R. Lavery, and J. H. Maddocks. Kinking occurs during molecular dynamics simulations of small DNA minicircles. *Structure*, 14:1527–1534, 2006.

[19] L. D. Landau and E. M. Lifshitz. *Theory of Elasticity*, volume 7. Elsevier, 3rd edition, 1986.

[20] O. C. Lee, C. Kim, J. Y. Kim, N. K. Lee, and W. Sung. Two conformational states in D-shaped DNA: Effects of local denaturation. *Sci. Rep.*, 6:28239, 2016.

[21] N. R. Markham and M. Zuker. DINAMelt web server for nucleic acid melting prediction. *Nucleic Acids Res.*, 33:W577–W581, 2005.

[22] J. F. Marko and E. D. Siggia. Stretching DNA. *Macromolecules*, 28:8759–8770, 1995.

[23] D. H. Mathews, J. Sabina, M. Zuker, and D. H. Turner. Expanded sequence dependence of thermodynamic parameters improves prediction of RNA secondary structure. *J. Mol. Biol.*, 288:911–940, 1999.

[24] A. K. Mazur and M. Maaloum. DNA flexibility on short length scales probed by atomic force microscopy. *Phys. Rev. Lett.*, 112:068104, 2014.

[25] C. V. Miduturu and S. K. Silverman. DNA constraints allow rational control of macromolecular conformation. *J. Am. Chem. Soc.*, 127:10144–10145, 2005.

[26] D. R. Nelson. Statistical physics of unzipping DNA. arXiv:cond-mat/0309559v1, 2003.

[27] M. S. Nepal, A. Yaniv, E. Shafran, and O. Krichevsky. Structure of DNA coils in dilute and semidilute solutions. *Phys. Rev. Lett.*, 110:T91, 2013.

[28] M. Peyrard, S. Cuesta-López, and G. James. Modelling DNA at the mesoscale: A challenge for nonlinear science? *Nonlinearity*, 21:T91, 2008.

[29] H. Qu, C.-Y. Tseng, Y. Wang, A. J. Levine, and G. Zocchi. The elastic energy of sharply bent nicked DNA. *Europhys. Lett.*, 90:10803, 2010.

[30] H. Qu, Y. Wang, C.-Y. Tseng, and G. Zocchi. Critical torque for kink formation in double-stranded DNA. *Phys. Rev. X*, 1:021008, 2011.

[31] H. Qu and G. Zocchi. The complete bending energy function for nicked DNA. *Europhys. Lett.*, 94:10803, 2011.

[32] D. Sanchez, H. Qu, D. Bulla, and G. Zocchi. DNA kinks and bubbles: Temperature dependence of the elastic energy of sharply bent 10-nm-size DNA molecules. *Phys. Rev. E*, 87:022710, 2013.

[33] E. Shafran, A. Yaniv, and O. Krichevsky. Marginal nature of DNA solutions. *Phys. Rev. Lett.*, 104:128101, 2010.

[34] R. Shusterman, S. Alon, T. Gavrinyov, and O. Krichevsky. Monomer dynamics in double- and single-stranded DNA polymers. *Phys. Rev. Lett.*, 92:048303, 2004.

[35] S. B. Smith, Y. Cui, and C. Bustamante. Overstretching B-DNA: The elastic response of individual double-stranded and single-stranded DNA molecules. *Science*, 271:795–799, 1996.

[36] K. Sneppen and G. Zocchi. *Physics in Molecular Biology*. Cambridge University Press, 2005.

[37] T. Strick, J.-F. Allemand, V. Croquette, and D. Bensimon. Twisting and stretching single DNA molecules. *Prog. Biophys. Mol. Biol.*, 74:115–140, 2000.

[38] A. Wang and G. Zocchi. Elastic energy driven polymerization. *Biophys. J.*, 96:2344–2352, 2009.

[39] P. A. Wiggins, T. van der Heijden, F. Moreno-Herrero, A. Spakowitz, R. Phillips, J. Widom, C. Dekker, and P. C. Nelson. High flexibility of DNA on short length scales probed by atomic force microscopy. *Nat. Nanotechnol.*, 1:137–141, 2006.

[40] P. A. Wiggins, R. Phillips, and P. C. Nelson. Exact theory of kinkable elastic polymers. *Phys. Rev. E*, 71:021909, 2005.

[41] J. Yan, R. Kawamura, and J. F. Marko. Statistics of loop formation along double helix DNAs. *Phys. Rev. E*, 71:061905, 2005.

[42] J. Yan and J. F. Marko. Localized single-stranded bubble mechanism for cyclization of short double helix DNA. *Phys. Rev. Lett.*, 93:108108, 2004.

[43] C. Yuan, H. Chen, X. W. Lou, and L. A. Archer. DNA bending stiffness on small length scales. *Phys. Rev. Lett.*, 100:018102, 2008.

[44] Y. Zeng, A. Montrichok, and G. Zocchi. Length and statistical weight of bubbles in DNA melting. *Phys. Rev. Lett.*, 91:148101, 2003.

[45] Y. Zeng, A. Montrichok, and G. Zocchi. Bubble nucleation and cooperativity in DNA melting. *J. Mol. Biol.*, 339:67–75, 2004.

[46] Y. Zhang and D. M. Crothers. High-throughput approach for detection of DNA bending and flexibility based on cyclization. *Proc. Natl. Acad. Sci. USA*, 100:3161–3166, 2003.

Chapter 3. Kinematics of Enzyme Action

[1] R. Afrin, M. T. Alam, and A. Ikai. Pretransition and progressive softening of bovine carbonic anhydrase II as probed by single molecule atomic force microscopy. *Protein Sci.*, 14:1447–1457, 2005.

[2] R. H. Austin, K. W. Beeson, L. Eisenstein, and H. Frauenfelder. Dynamics of ligand binding to myoglobin. *Biochemistry*, 14:5355, 1975.

[3] W. S. Bennett Jr and T. A. Steitz. Glucose-induced conformational change in yeast hexokinase. *Proc. Natl. Acad. Sci. USA*, 75:4848–4852, 1978.

[4] J. Blaszczyk, Y. Li, H. Yan, and X. Ji. Crystal structure of unligated guanylate kinase from yeast reveals GMP-induced conformational changes. *J. Mol. Biol.*, 307:247, 2001.

[5] L. A. Blumenfeld. The physical aspects of enzyme functioning. *J. Theor. Biol.*, 58:269, 1976.

[6] L. A. Blumenfeld and A. N. Tikhonov. *Thermodynamics of Intracellular Processes: Molecular Machines of the Living Cell*. Springer, New York, 1994.

[7] Y. Chen, S. E. Radford, and D. J. Brockwell. Force-induced remodelling of proteins and their complexes. *Curr. Opin. Struct. Biol.*, 30:89–99, 2015.

[8] B. Choi and G. Zocchi. Guanylate kinase, induced fit, and the allosteric spring probe. *Biophys. J.*, 92:1651–1658, 2007.

[9] B. Choi, G. Zocchi, Y. Wu, S. Chan, and L. J. Perry. Allosteric control through mechanical tension. *Phys. Rev. Lett.*, 95:078102, 2005.

[10] A. Cooper and D. T. Dryden. Allostery without conformational change. A plausible model. *Eur. Biophys. J.*, 11:103, 1984.

[11] T. Dauxois. Fermi, Pasta, Ulam, and a mysterious lady. *Phys. Today*, 6:55–57, 2008.

[12] O. Delalande, S. Sacquin-Mora, and M. Baaden. Enzyme closure and nucleotide binding structurally lock guanylate kinase. *Biophys. J.*, 101:1440–1449, 2011.

[13] H. Dietz, F. Berkemeier, M. Bertz, and M. Rief. Anisotropic deformation response of single protein molecules. *Proc. Natl. Acad. Sci. USA*, 103:12724–12728, 2006.

[14] W. Engelen, B. M. G. Janssen, and M. Merkx. DNA-based control of protein activity. *Chem. Commun.*, 52:3598, 2015.

[15] E. Evans and K. Ritchie. Strength of a weak bond connecting flexible polymer chains. *Biophys. J.*, 76:2439–2447, 1999.

[16] E. Fermi. *Thermodynamics*. Dover, 1937.

[17] A. Fersht. *Structure and Mechanism in Protein Science: A Guide to Enzyme Catalysis and Protein Folding*. W. H. Freeman, 1998.

[18] H. Frauenfelder, G. A. Petsko, and D. Tsernoglou. Temperature-dependent X-ray-diffraction as a probe of protein structural dynamics. *Nature*, 280:558, 1979.

[19] H. Frauenfelder, F. Parak, and R. Young. Conformational substates in proteins. *Annu. Rev. Biophys. Biophys. Chem.*, 17:451–479, 1988.

[20] H. Frauenfelder, B. H. McMahon, R. H. Austin, K. Chu, and J. T. Groves. The role of structure, energy landscape, dynamics, and allostery in the enzymatic function of myoglobin. *Proc. Natl. Acad. Sci. USA*, 98:2370–2374, 2001.

[21] P.-G. de Gennes. *Introduction to Polymer Dynamics*. Cambridge University Press, 1990.

[22] R. J. Hawkins and T. C. B. McLeish. Coarse-grained model of entropic allostery. *Phys. Rev. Lett.*, 93:098104, 2004.

[23] J. J. Hopfield. Relation between structure, co-operativity and spectra in a model of hemoglobin action. *J. Mol. Biol.*, 77:207–222, 1972.

[24] D. Kern and E. R. Zuiderweg. The role of dynamics in allosteric regulation. *Curr. Opin. Struct. Biol.*, 13:748–757, 2003.

[25] C. Kittel. *Elementary Statistical Physics*. Dover, 1990.

[26] A. B. Kolomeisky. Michaelis-Menten relations for complex enzymatic networks. *J. Chem. Phys.*, 134:155101, 2011.

[27] D. E. Koshland Jr. Application of a theory of enzyme specificity to protein synthesis. *Proc. Natl. Acad. Sci. USA*, 44:98–104, 1958.

[28] H. A. Kramers. Brownian motion in a field of force and the diffusion model of chemical reactions. *Physica*, 7:284–304, 1940.

[29] L. D. Landau and E. M. Lifshitz. *Theory of Elasticity*, volume 7. Elsevier, 3rd edition, 1986.

[30] Y. Li, Y. Zhang, and H. Yan. Kinetic and thermodynamic characterizations of yeast guanylate kinase. *J. Biol. Chem.*, 271:28038–28044, 1996.

[31] J. A. McCammon, B. R. Gelin, and M. Karplus. Dynamics of folded proteins. *Nature*, 267:585, 1977.

[32] R. Merkel, P. Nassoy, A. Leung, K. Ritchie, and E. Evans. Energy landscapes of receptor-ligand bonds explored with dynamic force spectroscopy. *Nature*, 397:50–53, 1998.

[33] J. Monod, J.-P. Changeux, and F. Jacob. Allosteric proteins and cellular control systems. *J. Mol. Biol.*, 6:306–329, 1963.

[34] T. E. Ouldridge. DNA nanotechnology: Understanding and optimisation through simulation. *Mol. Phys.*, 113:1–15, 2015.

[35] M. F. Perutz. Nature of haem-haem interaction. *Nature*, 237:495–499, 1972.

[36] M. Radmacher, M. Fritz, J. P. Cleveland, D. A. Walters, and P. K. Hansma. Imaging adhesion forces and elasticity of lysozyme adsorbed on mica with the atomic force microscope. *Langmuir*, 10:3809–3814, 1994.

[37] S. Sacquin-Mora, O. Delalande, and M. Baaden. Functional modes and residue flexibility control the anisotropic response of guanylate kinase to mechanical stress. *Biophys. J.*, 99:3412–3419, 2010.

[38] Y. Savir and T. Tlusty. Conformational proofreading: The impact of conformational changes on the specificity of molecular recognition. *PLoS ONE*, 2:e468, 2007.

[39] M. J. Schnitzer, K. Visscher, and S. M. Block. Force production by single kinesin motors. *Nat. Cell Biol.*, 2:718–723, 2000.

[40] K. Sneppen and G. Zocchi. *Physics in Molecular Biology*. Cambridge University Press, 2005.

[41] K. Svoboda and S. M. Block. Force and velocity measured for single kinesin molecules. *Cell*, 77:773–784, 1994.

[42] K. Svoboda, C. F. Schmidt, B. J. Schnapp, and S. M. Block. Direct observation of kinesin stepping by optical trapping interferometry. *Nature*, 365:721–727, 1993.

[43] C.-Y. Tseng, A. Wang, and G. Zocchi. Mechano-chemistry of the enzyme guanylate kinase. *Europhys. Lett.*, 91:18005, 2010.

[44] C.-Y. Tseng and G. Zocchi. Mechanical control of renilla luciferase. *J. Am. Chem. Soc.*, 135:11879, 2013.

[45] C.-Y. Tseng and G. Zocchi. Equilibrium softening of an enzyme explored with the DNA spring. *Appl. Phys. Lett.*, 104:153702, 2014.

[46] K. Visscher, M. J. Schnitzer, and S. M. Block. Single kinesin molecules studied with a molecular force clamp. *Nature*, 400:184–189, 1999.

[47] G. Zaccai. How soft is a protein? A protein dynamics force constant measured by neutron scattering. *Science*, 288:1604–1607, 2000.

[48] F. Zhang, J. Nangreave, Y. Liu, and H. Yan. Structural DNA nanotechnology: State of the art and future perspective. *J. Am. Chem. Soc.*, 136:11198, 1999.

[49] G. Zocchi. Controlling proteins through molecular springs. *Ann. Rev. Biophys.*, 38:75–88, 2009.

Chapter 4. Dynamics of Enzyme Action

[1] M. V. Akhterov et al. Observing lysozyme's closing and opening motions by high-resolution single-molecule enzymology. *ACS Chem. Biol.*, 10:1495, 2015.

[2] R. Afrin, M. T. Alam, and A. Ikai. Pretransition and progressive softening of bovine carbonic anhydrase II as probed by single molecule atomic force microscopy. *Protein Sci.*, 14:1447, 2005.

[3] Z. Alavi, A. Ariyaratne, and G. Zocchi. Nano-rheology measurements reveal that the hydration layer of enzymes partially controls conformational dynamics. *Appl. Phys. Lett.*, 106:203702, 2015.

[4] A. Ariyaratne, C. Wu, C.-Y. Tseng, and G. Zocchi. Dissipative dynamics of enzymes. *Phys. Rev. Lett.*, 113:198101, 2014.

[5] A. Ariyaratne and G. Zocchi. Plasmon resonance enhanced mechanical detection of ligand binding. *Appl. Phys. Lett.*, 106:013702, 2015.

[6] M. N. Artyomov, A. Yu. Morozov, and A. B. Kolomeisky. Dynamics of molecular motors in reversible burnt-bridge models. *Condens. Matter Phys.*, 13:23801, 2010.

[7] R. H. Austin, K. W. Beeson, L. Eisenstein, and H. Frauenfelder. Dynamics of ligand binding to myoglobin. *Biochemistry*, 14:5355, 1975.

[8] W. S. Bennett Jr and T. A. Steitz. Glucose-induced conformational change in yeast hexokinase. *Proc. Natl. Acad. Sci. USA*, 75:4848–4852, 1978.

[9] J. Blaszczyk, Y. Li, H. Yan, and X. Ji. Crystal structure of unligated guanylate kinase from yeast reveals GMP-induced conformational changes. *J. Mol. Biol.*, 307:247, 2001.

[10] L. A. Blumenfeld. The physical aspects of enzyme functioning. *J. Theor. Biol.*, 58:269, 1976.

[11] L. A. Blumenfeld and A. N. Tikhonov. *Thermodynamics of Intracellular Processes: Molecular Machines of the Living Cell*. Springer, New York, 1994.

[12] B. Brooks and M. Karplus. Harmonic dynamics of proteins: Normal modes and fluctuations in bovine pancreatic trypsin inhibitor. *Proc. Natl. Acad. Sci. USA*, 80:6571, 1983.

[13] C. Bustamante, D. Keller, and G. Oster. The physics of molecular motors. *Acc. Chem. Res.*, 34:412–420, 2001.

[14] A. Cooper and D. T. Dryden. Allostery without conformational change. A plausible model. *Eur. Biophys. J.*, 11:103, 1984.

[15] O. Delalande, S. Sacquin-Mora, and M. Baaden. Enzyme closure and nucleotide binding structurally lock guanylate kinase. *Biophys. J.*, 101:1440–1449, 2011.

[16] M. Dong, S. Husale, and O. Sahin. Determination of protein structural flexibility by microsecond force spectroscopy. *Nat. Nanotechnol.*, 4:514–517, 2009.

[17] E. Evans and K. Ritchie. Strength of a weak bond connecting flexible polymer chains. *Biophys. J.*, 76:2439, 1999.

[18] M. E. Fisher and A. B. Kolomeisky. The force exerted by a molecular motor. *Proc. Natl. Acad. Sci. USA*, 96:6597–6602, 1999.

[19] C. Fogle, J. Rudnick, and D. Jasnow. Protein viscoelastic dynamics: A model system. *Phys. Rev. E*, 92:032719, 2015.

[20] H. Frauenfelder, B. H. McMahon, R. H. Austin, K. Chu, and J. T. Groves. The role of structure, energy landscape, dynamics, and allostery in the enzymatic function of myoglobin. *Proc. Natl. Acad. Sci. USA*, 98:2370–2374, 2001.

[21] H. Frauenfelder, F. Parak, and R. Young. Conformational substates in proteins. *Annu. Rev. Biophys. Biophys. Chem.*, 17:451–479, 1988.

[22] H. Frauenfelder, G. A. Petsko, and D. Tsernoglou. Temperature-dependent X-ray-diffraction as a probe of protein structural dynamics. *Nature*, 280:558–563, 1979.

[23] N. Go, T. Noguti, and T. Nishikawa. Dynamics of a small globular protein in terms of low-frequency vibrational modes. *Proc. Natl. Acad. Sci. USA*, 80:3696–3700, 1983.

[24] P. Hammarström, R. Owenius, L. G. Mårtensson, U. Carlsson, and M. Lindgren. High-resolution probing of local conformational changes in proteins by the use of multiple labeling: Unfolding and self-assembly of human carbonic anhydrase II monitored by spin, fluorescent, and chemical reactivity probes. *Biophys. J.*, 80:2867–2885, 2001.

[25] R. J. Hawkins and T. C. B. McLeish. Coarse-grained model of entropic allostery. *Phys. Rev. Lett.*, 93:098104, 2004.

[26] J. J. Hopfield. Relation between structure, co-operativity and spectra in a model of hemoglobin action. *J. Mol. Biol.*, 77:207–222, 1972.

[27] J. Howard. The movement of kinesin along microtubules. *Ann. Rev. Physiol.*, 58:703–729, 1996.

[28] H. Jensenius and G. Zocchi. Measuring the spring constant of a single polymer chain. *Phys. Rev. Lett.*, 79:5030–5033, 1997.

[29] F. Julicher, A. Ajdari, and J. Prost. Modeling molecular motors. *Rev. Mod. Phys.*, 69:1269, 1997.

[30] D. Kern and E. R. Zuiderweg. The role of dynamics in allosteric regulation. *Curr. Opin. Struct. Biol.*, 13:748–757, 2003.

[31] D. E. Koshland Jr. Application of a theory of enzyme specificity to protein synthesis. *Proc. Natl. Acad. Sci. USA*, 44:98–104, 1958.

[32] H. A. Kramers. Brownian motion in a field of force and the diffusion model of chemical reactions. *Physica*, 7:284–304, 1940.

[33] L. D. Landau and E. M. Lifshitz. *Theory of Elasticity*, volume 7. Elsevier, 3rd edition, 1986.

[34] A. Langer et al. Protein analysis by time-resolved measurements with an electro-switchable DNA chip. *Nat. Commun.*, 4:2099, 2013.

[35] Y. Li, Y. Zhang, and H. Yan. Kinetic and thermodynamic characterizations of yeast guanylate kinase. *J. Biol. Chem.*, 271:28038, 1996.

[36] M. O. Magnasco. Forced thermal ratchets. *Phys. Rev. Lett.*, 71:1477, 1993.

[37] J. A. McCammon, B. R. Gelin, and M. Karplus. Dynamics of folded proteins. *Nature*, 267:585, 1977.

[38] R. Merkel, P. Nassoy, A. Leung, K. Ritchie, and E. Evans. Energy landscapes of receptor-ligand bonds explored with dynamic force spectroscopy. *Nature*, 397:50–53, 1998.

[39] M. R. Mitchell, T. Tlusty, and S. Leibler. Strain analysis of protein structures and low dimensionality of mechanical allosteric couplings. *Proc. Natl. Acad. Sci. USA*, 113:E5847–E5855, 2016.

[40] O. Miyashita, J. N. Onuchic, and P. G. Wolynes. Nonlinear elasticity, proteinquakes, and the energy landscapes of functional transitions in proteins. *Proc. Natl. Acad. Sci. USA*, 100:12570, 2003.

[41] J. J. Moc, R. T. Hill, Yu.-Ju. Tsai, A. Chilkoti, and D. R. Smith. Probing dynamically tunable localized surface plasmon resonances of film-coupled nanoparticles by evanescent wave excitation. *Nano Lett.*, 12:1757–1764, 2012.

[42] J. Monod, J.-P. Changeux, and F. Jacob. Allosteric proteins and cellular control systems. *J. Mol. Biol.*, 6:306–329, 1963.

[43] H. Noji, R. Yasuda, M. Yoshida, and K. Kinosita Jr. Direct observation of the rotation of F1 ATPase. *Nature*, 386:299–302, 1997.

[44] M. F. Perutz. Nature of haem-haem interaction. *Nature*, 237:495–499, 1972.

[45] K. W. Plaxco and D. Baker. Limited internal friction in the rate-limiting step of a two-state protein folding reaction. *Proc. Natl. Acad. Sci. USA*, 95:13591, 1998.

[46] H. Qian. Single-particle tracking: Brownian dynamics of viscoelastic materials. *Biophys. J.*, 79:137, 2000.

[47] H. Qian. The mathematical theory of molecular motor movement and chemomechanical energy transduction. *J. Math. Chem.*, 27:219, 2000.

[48] H. Qian, S. Kjelstrup, A. B. Kolomeisky, and D. Bedeaux. Entropy production in mesoscopic stochastic thermodynamics: Nonequilibrium kinetic cycles driven by chemical potentials, temperatures, and mechanical forces. *J. Phys. Condens. Matter*, 28:153004, 2016.

[49] H. Qu, J. Landy, and G. Zocchi. Cracking phase diagram for the dynamics of an enzyme. *Phys. Rev. E*, 86:041915, 2012.

[50] M. Radmacher, M. Fritz, J. P. Cleveland, D. A. Walters, and P. K. Hansma. Imaging adhesion forces and elasticity of lysozyme adsorbed on mica with the atomic force microscope. *Langmuir*, 10:3809, 1994.

[51] Y. Savir and T. Tlusty. Conformational proofreading: The impact of conformational changes on the specificity of molecular recognition. *PLoS ONE*, 2:e468, 2007.

[52] T. Schmiedl and U. Seifert. Optimal finite-time processes in stochastic thermodynamics. *Phys. Rev. Lett.*, 98:108301, 2007.

[53] M. J. Schnitzer and S. M. Block. Kinesin hydrolyses one ATP per 8-nm step. *Nature*, 388:386–390, 1997.

[54] D. A. Sivak and G. E. Crooks. Thermodynamic metrics and optimal paths. *Phys. Rev. Lett.*, 108:190602, 2012.

[55] R. Soheilifard, D. E. Makarov, and G. J. Rodin. Rigorous coarse-graining for the dynamics of linear systems with applications to relaxation dynamics in proteins. *J. Chem. Phys.*, 135:054107, 2011.

[56] K. Sneppen and G. Zocchi. *Physics in Molecular Biology*. Cambridge University Press, 2005.

[57] K. Svoboda and S. M. Block. Force and velocity measured for single kinesin molecules. *Cell*, 77:773–784, 1994.

[58]   T. Tlusty, A. Libchaber, and J.-P. Eckmann. Mapping sequence to mechanics: Proteins as learning amorphous matter. *Phys. Rev. X*, 7:021037, 2017.

[59]   C.-Y. Tseng, A. Wang, and G. Zocchi. Mechano-chemistry of the enzyme guanylate kinase. *Europhys. Lett.*, 91:18005, 2010.

[60]   K. Visscher, M. J. Schnitzer, and S. M. Block. Force production by single kinesin motors. *Nature Cell Biology*, 2:718–723, 2000.

[61]   Y. Wang and G. Zocchi. Elasticity of globular proteins measured from the ac susceptibility. *Phys. Rev. Lett.*, 105:238104, 2010.

[62]   Y. Wang and G. Zocchi. The folded protein as a viscoelastic solid. *Europhys. Lett.*, 96:18003, 2011.

[63]   Y. Wang and G. Zocchi. Viscoelastic transition and yield strain of the folded protein. *PLoS ONE*, 6:e28097, 2011.

[64]   P. C. Whitford, O. Miyashita, Y. Levy, and J. N. Onuchic. Conformational transitions of adenylate kinase: Switching by cracking. *J. Mol. Biol.*, 366:1661–1671, 2007.

[65]   I. Yoon et al. Nanofiber near-field light-matter interactions for enhanced detection of molecular level displacements and dynamics. *Nano Lett.*, 13:1440–1445, 2013.

[66]   G. Zaccai. How soft is a protein? A protein dynamics force constant measured by neutron scattering. *Science*, 288:1604, 2000.

[67]   M. Zacharias. Rapid protein-ligand docking using soft modes from molecular dynamics simulations to account for protein deformability: Binding of FK506 to FKBP. *Proteins: Struct., Funct., Bioinf.*, 54:759–767, 2004.

[68]   G. Zocchi. Controlling proteins through molecular springs. *Ann. Rev. Biophys.*, 38:75–88, 2009.

[69]   P. R. Zulkowski, D. Sivak, G. E. Crooks, and M. R. DeWeese. Geometry of thermodynamic control. *Phys. Rev. E*, 86:041148, 2012.

INDEX

Bending modulus, 56, 57
Bubble states, 40

Central limit theorem, 3
Chemical potential, 31, 66, 83, 85
Chemo-dynamic cycle, 89
Conformational states, 153, 155
Cooperativity: and DNA melting, 34, 46; and
    hierarchical Hamiltonian, 46; and local
    rules, 47
Correlation function, 14, 57
Critical bending torque, 73, 99, 100, 119
Current in barrier crossing, 17, 18

D-DNA, 65
Diffusion, 1; constant, 2, 11, 92, 140;
    equation, 11, 91
Dissipation parameter, 8, 15, 122, 137
DNA: dimer states, 48; kink, 68, 73; melting
    curves, 53; nanorod, 27, 65, 89; spring,
    90, 97; structure, 26

Einstein relation, 12, 92, 159
Elastic energy: of D-DNA, 67; of DNA
    nanorod, 77; of enzyme-DNA chimera,
    104; and Euler instability, 68, 69
End-to-end distance, 2, 63, 69, 74, 100, 118;
    probability distribution of, 3, 7
Energy landscape, 157
Entropy: of dissociation, 32, 33, 39; and
    mechano-chemical coupling 112, 138;
    and nonlinearity, 55; and observation
    time, 23;
Enzyme-DNA chimera, 91, 98
Equilibrium: in statistical mechanics, 22
Error function, 107

Fluctuation-dissipation theorem, 15
Fokker-Planck equation, 159
Force-velocity curve, 102, 104

Gaussian network, 150
Gel electrophoresis, 38, 67
Golden rule, 83

Hydrogen bond, 26

Induced-fit, 94

Kinesin, 102, 103, 136, 137
Kuhn length, 4, 72

Langevin equation, 8, 15, 158
Law of mass action, 30

Melting curve, 38, 39
Melting temperature: concentration
    dependence, 32
Michaelis-Menten constant, 83, 88, 95, 102
Minimum work protocol, 161
Mobility, 8, 92, 128, 159

Optical trap, 102

Partition sum: enzyme-DNA chimera , 118;
    harmonic oscillator, 138; Michaelis-
    Menten kinetics, 84; Nearest Neighbor
    model, 54; 2x2 model, 51; Zipper model,
    35, 54
PBD model, 55
Persistence length, 57; and bending
    modulus, 58
Product inhibition, 84

Quenching method, 37

Random walk, 1
Rate: equations, 87, 147; of escape, 22, 127;
    of mechanical motion, 133

Softening transition: and D-DNA, 67; and
    DNA overstretching, 64; and enzyme
    mechanics, 118, 126, 143
Soft modes, 153
Stall force, 102, 134, 141

Transfer matrix, 48

Two-states system, 23; and DNA melting, 29; and enzymatic activity, 115; partition sum, 33

Unstacking transition, 38, 44

Viscoelasticity, 123
Viscosity, 22, 92, 137

Worm-like-chain, 60; Force-extension, 62; Hamiltonian, 60; Marko-Siggia formula, 63

Young modulus, 56, 69, 137

Zipper model: Hamiltonian, 46; Partition sum, 33

A NOTE ON THE TYPE

This book has been composed in Adobe Text and Gotham.
Adobe Text, designed by Robert Slimbach for Adobe,
bridges the gap between fifteenth- and sixteenth-century
calligraphic and eighteenth-century Modern styles.
Gotham, inspired by New York street signs, was designed
by Tobias Frere-Jones for Hoefler & Co.